The AI Delusion

THE AI DELUSION

GARY SMITH

Whatever ⟶ Nonsenes

OXFORD
UNIVERSITY PRESS

Great Clarendon Street, Oxford, OX2 6DP,
United Kingdom

Oxford University Press is a department of the University of Oxford.
It furthers the University's objective of excellence in research, scholarship,
and education by publishing worldwide. Oxford is a registered trade mark of
Oxford University Press in the UK and in certain other countries

First Edition published in 2018

Impression: 1

Published in the United States of America by Oxford University Press
198 Madison Avenue, New York, NY 10016, United States of America

British Library Cataloguing in Publication
Data available

Library of Congress Control Number: 2018933899

ISBN 978-0-19-882430-5

Printed and bound by
CPI Group (UK) Ltd, Croydon, CR0 4YY

CONTENTS

Introduction

The Democratic Party's 2008 presidential nomination was supposed to be the inevitable coronation of Hillary Clinton. She was the most well-known candidate; had the most support from the party establishment, and had, by far, the most financial resources.

Two big names (Al Gore and John Kerry) considered running, but decided they had no hope of defeating the Clinton machine. That left an unlikely assortment of lesser-knowns: a U.S. Representative from Ohio (Dennis Kucinich), the Governor of New Mexico (Bill Richardson), and several U.S. Senators: Joe Biden (Delaware), John Edwards (North Carolina), Chris Dodd (Connecticut), Mike Gravel (Alaska), and Barack Obama (Illinois).

The nomination went off script. Obama was a first-term senator, a black man with an unhelpful name, but he excited voters. He raised enough money to be competitive in the Iowa caucuses and he persuaded Oprah Winfrey to campaign for him. Obama defeated Clinton by eight percentage points in Iowa and the race was on.

Obama won the Democratic nomination and, then, the presidential election against Republican John McCain because the Obama campaign had a lot more going for it than Obama's eloquence and charisma: Big Data.

The Obama campaign tried to put every potential voter into its data base, along with hundreds of tidbits of personal information: age, gender, marital status, race, religion, address, occupation, income, car registrations, home value, donation history, magazine subscriptions, leisure activities, Facebook friends, and anything else they could find that seemed relevant.

Some data were collected from public data bases, some from e-mail exchanges or campaign workers knocking on front doors. Some data were purchased from private data vendors. Layered on top were weekly telephone surveys of thousands of potential voters which not only gathered personal data, but also attempted to gauge each person's likelihood of voting—and voting for Obama.

These voter likelihoods were correlated statistically with personal characteristics and extrapolated to other potential voters based on their personal characteristics. The campaign's computer software predicted how likely each person its data base was to vote and the probability that the vote would be for Obama.

This data-driven model allowed the campaign to microtarget individuals through e-mails, snail mail, personal visits, and television ads asking for donations and votes. If the computer program predicted that people with hunting licenses were likely to be opposed to gun-control legislation, then gun-control was less likely to be mentioned in pitches to people with hunting licenses. The software suggested other levers that could be used to secure donations and votes.

In the crucial month of January, 2008, Obama raised $36 million, a record for any politician, and nearly three times the amount raised by Clinton. After Obama secured the nomination, the fund-raising continued. For the full 2008 election campaign, Obama raised $780 million, more than twice the amount raised by his Republican opponent, John McCain. McCain didn't have a realistic chance of winning, and he didn't—with only 173 electoral votes to Obama's 365.

Eight years later, Hillary Clinton made another presidential run, determined to have Big Data on her side.

This time, Big Data failed.

The Clinton campaign hired 60 mathematicians and statisticians, several from the Obama campaign, to create a software program that was named Ada in honor of a 19th-century female mathematician, Ada, Countess of Lovelace. After Clinton became the first female president, she would reveal Ada to be the secret behind her success. What a great story!

Ada was housed on its own server with access restricted to a handful of people. Some knew that there was a "model." but they had no idea how it worked. Most knew nothing at all.

On September 16, 2016, seven weeks before the election, Eric Siegel wrote an article in *Scientific American* titled, "How Hillary's Campaign Is (Almost Certainly) Using Big Data." He argued that, "The evidence suggests her campaign is using a highly targeted technique that worked for Obama." A year and a half into the campaign, observers were still speculating on Clinton's use of Big Data. That's how carefully Ada had been hidden.

The Clinton campaign was extremely secretive about Ada—certainly because they did not want to give Clinton's opponents any ideas, and perhaps because they didn't want to fuel the stereotype that the Clinton campaign was mechanical, cautious, and scripted—without the inspirational passion that Bernie Sanders and Donald Trump brought to their campaigns.

Ada ran 400,000 simulations a day predicting the election outcome for scenarios that it considered plausible. What if the turnout in Florida was two-percentage-points higher and the turnout in New Mexico was one-percentage-point lower? What if...? What if...? The results were summarized, most importantly by identifying geographic areas where resources should be deployed and which resources should be used.

For example, 70 percent of the campaign budget went for television ads, and Ada determined virtually every dollar spent on these ads. The advice of experienced media advisors was neither sought nor heeded. Ada's data base contained detailed socioeconomic information on which people watched which television shows in which cities, and Ada estimated how likely they were to vote for Clinton. Ada used these data to calculate the theoretical cost of every potential vote and to determine how much money to spend on ads on different shows, at different times, and in different television markets.

No one really knew exactly how Ada made her decisions, but they did know that she was a powerful computer program analyzing an unimaginable amount of data. So, they trusted her. She was like an omniscient goddess. Don't ask questions, just listen.

We still do not know how Ada determined what she considered to be an optimal strategy, but it is clear that, based on historical data, Ada took blue-collar voters for granted, figuring that they reliably voted Democratic, most recently for Obama, and they would do so again. With blue-collar votes as her unshakeable base, Clinton would coast to victory by ensuring that minorities and liberal elites turned out to vote for her. This presumption was exacerbated by Ada's decision that the campaign did not need to spend money doing polling in safe states—so, the campaign did not realize that some safe states were no longer safe until it was too late.

Ada is just a computer program and, like all computer programs, has no common sense or wisdom. Any human who had been paying the slightest bit of attention noticed Clinton's vulnerability against Bernie Sanders, a virtually unknown 74-year-old Socialist senator from Vermont, who was not even a Democrat until he decided to challenge Clinton. A human would have tried to figure out why Sanders was doing so well; Ada didn't.

When Clinton suffered a shock defeat to Sanders in the Michigan primary, it was obvious to people with campaign experience that his populist message had tremendous appeal, and that the blue-collar vote could not be taken for granted. Ada didn't notice.

Clinton was furious at being blind-sided in Michigan but she was still confident that Ada knew best. Clinton blamed her shock loss on everything but Ada. Ada was, after all, a powerful computer—free of human biases, churning through gigabytes of data, and producing an unimaginable 400,000 simulations a day. No human could compete with that. So, the campaign kept to its data-driven playbook, largely ignoring the pleas of seasoned political experts and campaign workers who were on the ground talking to real voters.

Ada did not compare the enthusiasm of the large crowds that turned out, first for Sanders, and later for Trump, to the relatively subdued small crowds that listened to Clinton. There were no enthusiasm numbers for Ada to crunch, so Ada ignored energy and passion, and Clinton's data-driven campaign did, too. To a computer, if it can't be measured, it isn't important.

Most glaringly, the Clinton campaign's data wonks shut out Bill Clinton, perhaps the best campaigner any of us have ever seen. The centerpiece of his successful 1992 election campaign against the incumbent president, George H. W. Bush, was "It's the economy, stupid." Bill instinctively knew what mattered to voters and how to persuade them that he cared.

In the 2016 election, Bill Clinton saw the excitement generated by Bernie Sanders and Donald Trump in their appeal to working-class voters and he counseled that "It's the economy, stupid" should be the defining issue of Hillary's campaign—particularly in the Midwestern rust-belt states of Ohio, Pennsylvania, Michigan, and Wisconsin, the so-called Blue

Wall, the firewall of reliably blue states that Ada assumed would be the base for Clinton's victory over Donald Trump.

Another Ada blind spot was that seasoned politicians know that television ads are okay, but rural voters are most impressed if candidates prove they care by taking time to show up at town hall meetings and county fairs. Going by the numbers, Ada did not know this. When in the waning days of the campaign, it was decided that *one* staffer should spend time on rural outreach, the staffer was from Brooklyn—not a promising background if you're looking for someone who can relate to farmers.

Bill was outraged that Hillary did not listen to him during the campaign—literally refusing to take his phone calls. He complained to Hillary's campaign chairman, John Podesta that, "Those snotty-nosed kids over there are blowing this thing because nobody is listening to me."

Ada concluded that voters were more worried about Trump's unpresidential behavior than they were about jobs; so Hillary focused her campaign on anti-Trump messages: "Hey, I may not be perfect, but Trump is worse."

Following Ada's advice, the Clinton campaign almost completely ignored Michigan and Wisconsin, even though her primary-campaign losses to Bernie Sanders in both states should have been a fire-alarm of a wake-up call. Instead, Clinton wasted time and resources campaigning in places like Arizona—states she probably would not win (and did not win)—because Ada decided that Clinton could secure a landslide victory with wins in marginally important states.

In the aftermath, a Democratic pollster said that, "It's nothing short of malpractice that her campaign didn't look at the electoral college and put substantial resources in states like Michigan and Wisconsin."

After the loss, Bill pointed his middle finger at the data wonks who put all their faith in a computer program and ignored the millions of working-class voters who had either lost their jobs or feared they might lose their jobs. In one phone call with Hillary, Bill reportedly got so angry that he threw his phone out the window of his Arkansas penthouse.

We don't know if it was bad data or a bad model, but we do know that Big Data is not a panacea—particularly when Big Data is hidden inside a computer and humans who know a lot about the real world do not know what the computer is doing with all that data.

Computers can do some things really, really well. We are empowered and enriched by them every single day of our lives. However, Hillary Clinton is not the only one who has been overawed by Big Data, and she will surely not be the last. My hope is that I can persuade you not to join their ranks.

CHAPTER 1

Intelligent or obedient?

J eopardy! is a popular game show that, in various incarnations, has been on television for more than 50 years. The show is a test of general knowledge with the twist that the clues are answers and the contestants respond with questions that fit the answers. For example, the clue, "16th President of the United States," would be answered correctly with "Who is Abraham Lincoln?" There are three contestants, and the first person to push his or her button is given the first chance to answer the question orally (with the exception of the Final Jeopardy clue, when all three contestants are given 30 seconds to write down their answers).

SINK IT AND
YOU'VE SCRATCHED

In many ways, the show is ideally suited for computers because computers can store and retrieve vast amounts of information without error. (At a teen Jeopardy tournament, a boy lost the championship because he wrote "Who is Annie Frank?" instead of "Who is Anne Frank." A computer would not make such an error.)

On the other hand, the clues are not always straightforward, and sometimes obscure. One clue was "Sink it and you've scratched." It is difficult for a computer that is nothing more than an encyclopedia of facts to come up with the correct answer: "What is the cue ball?"

Another challenging clue was, "When translated, the full name of this major league baseball team gets you a double redundancy." (Answer: "What is the Los Angeles Angels?")

In 2005 a team of 15 IBM engineers set out to design a computer that could compete with the best Jeopardy players. They named it Watson, after IBM's first CEO, Thomas J. Watson, who expanded IBM from 1,300 employees and less than $5 million in revenue in 1914 to 72,500 employees and $900 million in revenue when he died in 1956.

The Watson program stored the equivalent of 200 million pages of information and could process the equivalent of a million books per second. Beyond its massive memory and processing speed, Watson can understand natural spoken language and use synthesized speech to communicate. Unlike search engines that provide a list of relevant documents or web sites, Watson was programmed to find specific answers to clues.

Watson used hundreds of software programs to identify the keywords and phrases in a clue, match these to keywords and phrases in its massive data base, and then formulate possible responses. If the response is a name, like Abraham Lincoln, Watson was programmed to state a question starting with "Who is." For a thing, Watson starts with "What is." The more the individual software programs agree on an answer, the more certain the Watson program is that this is the correct answer.

Watson can answer straightforward clues like "The 16th President" easily, but struggles with words that have multiple meanings, like "Sink it and you've scratched." On the other hand, Watson does not get nervous and never forgets anything.

Watson was ready to take on Jeopardy in 2008, but there were issues to be negotiated. The IBM team was afraid that the Jeopardy staff would write clues with puns and double meanings that could trick Watson. That, in and of itself, reveals one big difference between humans and computers. Humans can appreciate puns, jokes, riddles, and sarcasm because we understand words in context. The best that current computers can do is check whether the pun, joke, riddle, or sarcastic comment has been stored in its data base.

The Jeopardy staff agreed to select clues randomly from a stockpile of clues that had been written in the past, but never used. On the other hand, the Jeopardy staff were afraid that if Watson emitted an electronic signal when it had an answer, it would have an advantage over human contestants who must press buttons. The IBM team agreed to give Watson an electronic finger to push a button, but it was still faster than humans, and gave Watson a decisive advantage. Is a fast trigger finger intelligence? How would the match have turned out if Watson's reaction time had been slowed to match that of humans?

In the man-versus-machine challenge in 2011, Watson played a two-round match against two popular former Jeopardy champions, Ken Jennings and Brad Rutter. In the first round, the Final Jeopardy clue was:

> *Its largest airport is named for a World War II hero;*
> *its second largest, for a World War II battle*

The two human contestants gave the correct answer, "What is Chicago?" Watson answered "What is Toronto?????," Watson evidently picked out the phrases *largest airport*, *World War II hero*, and *World War II battle*, and searched for common themes in its data base, not understanding that the second part of the clue ("*its second largest*") referred to the city's second largest airport. Watson added the multiple question marks because it calculated the probability of being correct at only 14 percent.

Nonetheless, Watson won easily with $77,147, compared to $24,000 for Jennings and $21,600 for Rutter. Watson received a $1 million prize for first place (which IBM donated to charity). Jennings and Rutter donated half of their respective prizes of $300,000 and $200,000 to charity.

Watson's Jeopardy triumph was a publicity bonanza that was worth millions. After its stunning victory, IBM announced that Watson's question-answering skills would be used for more important things than jousting with Alex Trebek, host of Jeopardy. IBM has been deploying Watson in health-care, banking, tech support, and other fields where massive data bases can be used to provide specific answers to specific questions.

For many people, Watson's defeat of these two great Jeopardy champions was proof beyond doubt that computers are smarter than humans. The mighty Watson knows all! If computers are now smarter than us, we should rely on them and trust their decisions. Maybe we should fear that they will soon enslave or exterminate us.

Is Watson really smarter than us? Watson's victory illustrates both the strengths and weaknesses of current computers. Watson is an astonishingly powerful search engine capable of finding words and phrases quickly in its massive data base (and it has a fast electronic trigger finger). However, I avoided using the word *read* because Watson does not know what words and phrases mean, like *World War II* and *Toronto*, nor does it understand words in context, like "*its second largest*". Watson's prowess is wildly exaggerated. Like many computer programs, Watson's seeming intelligence is just an illusion.

Watson's performance is in many ways a deception designed to make a very narrowly defined set of skills seem superhuman. Imagine a massive library with 200 million pages of English words and phrases and a human who does not understand English, but has an infinite amount of time to browse through this library looking for matching words and phrases. Would we say that this person is smart? Are computers super smart because they search for matches faster than humans can?

Even Dave Ferrucci, the head of IBM's Watson team, admitted, "Did we sit down when we built Watson and try to model human cognition? Absolutely not. We just tried to create a machine that could win at Jeopardy."

Board games

Computers have not only beaten humans at Jeopardy, they have defeated the best checkers, chess, and Go players, fueling the popular perception that computers are smarter than the smartest humans. These strategic board games require much, much more than a powerful search engine that matches words and phrases. The humans who excel at these games analyze board positions, formulate creative strategies, and plan ahead. Isn't that real intelligence?

Let's see. We'll start with a very simple childhood game.

Tic-tac-toe

In tic-tac-toe, two opponents take turns putting Xs and Os on a 3-by-3 grid. A player who gets three squares in a row—horizontally, vertically, or diagonally—has a tic-tac-toe and wins the game.

A software engineer could write a brute-force computer program to master tic-tac-toe by analyzing all possible sequences of moves. The first player has nine possible squares to choose from. For each of these possible first moves, the second player has eight choices, giving 72 first/second pairs. For each of these 72 pairs, the first player now has seven possible squares left. Overall, a complete game has $9 \times 8 \times 7 \times 6 \times 5 \times \times 3 \times 2 \times 1 =$

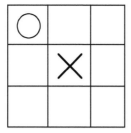

362, 880 possible sequences of choices that must be considered by the computer program.

There are more elegant ways to analyze all possible sequences, but the point is that tic-tac-toe programs do not look at the game the way humans do. Humans look at the 3-by-3 grid and think about which squares open up tic-tac-toe possibilities and which squares block the opponent's possible tic-tac-toes. A computer program does not visualize the squares. Instead, the programmer assigns each square a number, 1 through 9, and identifies the winning combinations (such as 1, 2, 3 and 1, 5, 9).

1	2	3
4	5	6
7	8	9

The computer program considers the possible sequences of the numbers 1 through 9 and identifies which strategies are optimal for each player, assuming that the opponent chooses optimal strategies. Once the software has been written and debugged, the best strategies are revealed instantly.

Assuming optimal play by the second player, the first player should begin with the center square or any of the four corner squares, and the second player should do the opposite, choosing a corner if the first player chose the center, and choosing the center if the first player chose a corner. With optimal play, the game always ends in a tie.

This is brute force computing in that no logical reasoning is involved, just a mindless enumeration of the permutations of the numbers 1 through 9 and the identification of the winning permutations.

In tic-tac-toe and other games, humans generally avoid brute force calculations of all possible move sequences because the number of possibilities explodes very quickly. Instead, we use logical reasoning to focus our attention on moves that make sense. Unlike a brute-force computer program, we don't waste time thinking through the implications of obviously wrong steps. Computers analyze stupid strategies because they do not have logic or common sense.

A human playing tic-tac-toe for the first time might study the 3-by-3 grid. While a computer plays around with the numbers 1 through 9, a human visualizes moves. She might immediately be attracted to the center square, recognizing that this square allows for four possible winning positions, compared to three for each of the corner squares, and two for each of the side squares.

The center square is also a great defensive move, in that any square chosen by the second player has, at most, only two possible winning positions. A first move in the corner or side, in contrast, allows the second player to seize the middle square, blocking one of the first player's winning positions while creating three possible winning positions for the second player.

Logically, the middle square seems to be the best opening move and a side square appears to be the least attractive. This human visualization of the board and identification of the strategic value of the center square is completely different from a software program's mindless consideration of all possible permutations of the numbers 1 through 9.

A human would also immediately recognize the symmetry of the game—each of the four corner squares is equally attractive (or unattractive) for an opening move. So, the human only has to think through the consequences of choosing one of the corners, and the same logic will apply to the other three corners. At every step, the symmetry of the game allows the human to reduce the number of possible moves being considered. Finally, the human will recognize that some moves are attractive because they force the opponent to choose a bad square in order to block an immediate tic-tac-toe.

With strategic thinking, a human can figure out the optimal strategy and recognize that optimally played games always end in draws. With experience, a human will learn that games played against children can sometimes be won by playing unconventionally; for example, opening with a corner square or even a side square.

Ironically, even though humans might use logic to figure out the optimal strategies, a computer software program written by humans might defeat humans because computers do *not* have to think about their moves. A computer tic-tac-toe program simply obeys the rules programmed into it. In contrast, humans must think before moving and will eventually get tired and make mistakes.

The advantage that computers have over humans has nothing to do with *intelligence* in the way the word is normally used. It is humans who write the software that identifies the optimal strategies and stores these strategies in the computer's memory so that the computer has rules to obey.

Even though tic-tac-toe is a children's game that is ultimately boring, it is a nice example because it highlights the power and limitations of computer software. Computer programs are very useful for tedious computations. Well-written software gives the same answer every time and never tires of doing what it has been programmed to do. Computers process faster and remember more than humans.

How can a human ever hope to compete with a computer? Certainly not in activities where memory and processing speed are all that matter. Perhaps the real miracle is not that computers are so powerful, but that there are still many things that humans do better than computers. Following rules is very different from the instinctive intelligence that humans acquire during their lifetimes.

Human intelligence allows us to recognize cryptic language and distorted images, to understand why things happen, to react to unusual events, and so much more that would be beyond our grasp if we were mere rules-followers.

Checkers

Checkers is much more complicated than tic-tac-toe, indeed so complicated that a brute-force analysis of all possible sequences of moves is not practical. So, you might think that computers would have to mimic human thinking in order to play well. Nope.

American checkers (also called English draughts) is played on an 8-by-8 checkerboard, with alternating dark and light squares. Only the dark squares are used, which reduces the number of playable squares from 64 to 32. Each player begins with 12 pieces, traditionally flat and cylindrical, like hockey pucks, placed on the dark squares in front of them, with the 8 middle dark squares left open. The pieces are moved diagonally along the dark squares and capture an opponent's piece by jumping over it.

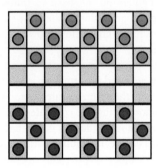

In theory, brute-force software could work though all the possible sequences for an unlimited number of moves ahead and identify the optimal strategies, just as in tic-tac-toe. However, there are too many possible sequences for current computers to analyze them all in a reasonable amount of time. So, humans have developed simplifying strategies for harnessing the power of computers. As with tic-tac-toe, computer checkers programs do not try to formulate logically appealing strategies. Instead, they exploit a computer's strengths: fast processing and perfect memories.

A tic-tac-toe game is over after nine moves. Checkers does not have a limited number of moves since players can move their pieces back and forth endlessly without either side winning. In practice, back-and-forth games are boring, so players agree to a draw when it is clear that, barring an idiotic blunder, neither player will ever win. (A ruthless checkers program would never agree to a draw, playing on until the human opponent is too tired to think clearly and makes a mistake.)

Although a checkers game has an unlimited number of moves, there are a fixed number of possible board positions. Instead of working out all possible sequences of moves, a more promising route for a checkers-playing computer is to look at all possible board positions and determine which moves from these positions are improvements and which are setbacks.

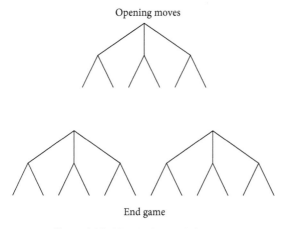

Opening moves

End game

Figure 1 Modeling checkers as decision trees

The task is still daunting. There are 500 billion billion possible board positions and one cannot truly determine whether a move is an improvement without looking at all possible sequences of positions that might follow.

The human insight is to break the game into three parts (opening moves, middle play, and end game), analyze each separately, and then link them.

For the beginning game, there are well-established "playbooks" that show the best possible opening moves, the best responses to every opening move, and so on, for several moves into the game. These playbooks are the collective wisdom that checkers players have accumulated over centuries of play. Every serious player studies these playbooks. A software engineer writing the code for a checkers program loads a playbook into the computer's memory, and the computer obeys these rules in the opening game.

For the end game, there are a relatively limited number of positions if there are only two pieces left on the board, a larger but still manageable number of positions for three pieces, and so on. For each of these possible positions, expert human checkers players can work out the optimal play and determine whether optimal play leads to a draw or a win for either player. As the number of remaining checkers pieces increases, the number of possible positions increases rapidly, but many are easily solved and the symmetry of the board reduces the number of positions that must be analyzed. After humans have analyzed all possible endings for all possible board positions involving, say fewer than six pieces, the optimal end-game play for each position is loaded into the computer's memory.

When a checkers game reaches one of these pre-loaded end-game positions, the computer obeys the rules by making the move that humans have identified as best. After the human makes a move in the end game of a computer-versus-human checkers game, the computer matches the new board position to one stored in its data base and makes the pre-determined optimal move. This continues until the games ends, usually by one side conceding or both sides agreeing to a draw.

A computer's middle game attempts to link the opening-move playbook with the end-game positions. If, after several opening moves, the play leads to a stored end-game position, the outcome of the game is known (assuming optimal play).

There are far too many possible middle-game positions for a brute force analysis to identify optimal sequences, so programmers combine human wisdom about checkers with the computer's power to enumerate sequences. If the computer has the power and time to look four moves ahead, then the computer looks at all possible four-move sequences and uses a human-specified loss function to compare all possible positions after four moves. The loss function, again based on centuries of human experience, takes into account factors that are thought to be important, like the number of pieces each player has and control of the center of the board. The human experts advising the programmer assign weights to the different factors to reflect the perceived importance of each.

The computer is typically programmed to choose the move that is *minmax*, in that it minimizes the possible loss (that's the *min*) in a worst-case scenario (that's the *max*). The program selects the move that has the smallest loss (or largest gain) if the other player follows an optimal strategy.

After several middle-game moves, the number of pieces shrinks to a level where the look-ahead calculations lead to the known end-game outcomes. Assuming optimal play, the game is essentially over. If a human player makes a mistake, the game may end sooner but it will not end better for the human player.

Notice how little "intelligence" is involved in computer checkers programs. The computer software obediently follows its opening-move instructions in the beginning and its end-game instructions at the end.

For the middle game, the computer software determines the look-ahead sequences and uses the human-specified loss function to determine its move, which it obediently makes.

Computer programs that are designed to play checkers, chess, Go, and other complex games do not attempt to mimic human thinking, which involves a creative recognition of the underlying principles that lead to victory. Instead, computer programs are built to exploit a computer's strengths—that it has an infallible memory and can obey rules flawlessly.

A checkers-playing computer program has several important advantages over human players. It never makes a mistake in its opening or ending moves. Human players may have studied a checkers playbook, but human memory is imperfect and humans may blunder. No human has considered, let along memorized, all possible end-game sequences, some of which require dozens of precise moves to reach the optimal outcome. A computer has the optimal sequences loaded into its data base; humans must figure out the optimal play on the fly and may err.

A human's only real chance to beat a computer checkers program is in the middle game. Humans may not be able to think as far ahead as a computer, analyzing as many possible sequences of moves, but some humans may have a better grasp of the strategic value of certain positions. For example, a human might recognize that controlling the center of the board is more important than the weights given by computer's loss function. Or the computer's numerical measure of the control of the center may be flawed. Or a human player may recognize that the ultimate control of the center depends on more than can be measured by looking at the current position.

The final advantage for a computer is that it does not get tired. A high-level checkers game can last for more than two hours. Because most games end in draws, a large number of games are played in checkers tournaments, perhaps four games a day for more than a week. Human players who must think about their moves for eight-to-ten hours a day, day after day, become fatigued and are prone to mistakes. Computers do not get tired, because they do not think. They obey.

The best checkers player in history was the legendary Marion Tinsley. He had been a precocious child, skipping four of his first eight years of school and becoming a mathematics professor specializing in combinatorial analysis. As a child, he studied checkers eight hours a day, five days a week. In graduate school, he claimed to have spent 10,000 hours studying checkers. In his twenties, he was virtually unbeatable.

Tinsley retired from tournament play for 12 years, reportedly because he was bored by opponents who played very conservatively, figuring that the best they could hope for was a draw. After returning to the game, he retired again in 1991, at age 63, but was lured back into the game by Jonathan Schaeffer, a mathematics professor who led a team that created Chinook, a computer checkers program. Schaeffer had three people on his research team—one specializing in the opening-move data base, one specializing in the end-game data base, and one responsible for the intermediate-game loss function.

In their 40-game match in 1992, most of the games were draws. Tinsley won game 5 when Chinook followed a line of play programmed into its playbook that was, in fact, suboptimal. Tinsley lost game 8, and attributed it to fatigue. Tinsley also lost game 14 when Chinook followed a sequence of moves in its data base that Tinsley had used years earlier, but forgotten. Game 18 went to Tinsley when Chinook malfunctioned (computer fatigue?). Tinsley then won games 25 and 39 and was declared the winner, 4 games to 2 with 33 draws.

It was a victory for man over machine, but it was also only the 6th and 7th games that Tinsley had lost in the thousands of tournament games he had played over a 45-year career.

Schaeffer enlarged Chinook's opening and end-game data bases enormously and increased its look-ahead capacity in the middle game from 17 moves to 19. He asked for a rematch in 1994. The first six games were draws and Tinsley recognized Chinook's improved performance. He said that he only had 10–12 moves to gain an advantage over Chinook before it got close enough to its enormous end-game data base so that it would not make any mistakes. Tragically, Tinsley had to abandon the match when it was discovered that he had pancreatic cancer. He died seven months later.

Tinsley had an incredible memory. After the first game of the 1992 match, he talked to Schaeffer about a game he had played more than 40 years ago, remembering every move perfectly. Still, his memory was no match for a powerful computer. What Tinsley did have was a feel for the game developed over many years of studying and playing checkers. There is no way that Chinook could have the same intuitive grasp of the strengths and weakness of positions.

In 14 exhibition games before their scheduled showdown, Tinsley and Chinook tied 13 games, with Tinsley securing the lone victory, in game 10. Schaeffer later wrote about that decisive game:

I reached out to play Chinook's 10th move. I no sooner released the piece when Tinsley looked up in surprise and said "You're going to regret that." Being inexperienced in the ways of the great Tinsley, I sat there silently thinking "What do you know? My program is searching 20 moves deep and says it has an advantage." Several moves later, Chinook's assessment dropped to equality. A few moves later, it said Tinsley was better. Later Chinook said it was in trouble. Finally, things became so bad we resigned. In his notes to the game, Tinsley revealed that he had seen to the end of the game and knew he was going to win on move 11, one move after our mistake. Chinook needed to look ahead 60 moves to know that its 10th move was a loser.

After Tinsley's death, Chinook played a 32-game match against Don Lafferty, the second best player in the world, and won 1–0 with 31 draws. In 1996, Chinook was retired from tournament play, although you can play an online game against a weakened version of Chinook. After its retirement, Chinook joined dozens of other computers that had been running more or less continuously for 18 years to determine whether a checkers player moving first and making optimal moves could guarantee a victory.

In 2007, Schaeffer announced that, like tic-tac-toe, checkers is a perfectly balanced game in that if each player plays optimally, a draw is guaranteed. This was a great computational feat, but I wouldn't call it intelligence.

The next generation of game-playing computer programs has taken a different route—a trial-and-error process in which a computer plays millions of games against itself and records what works. Using this approach, a program named AlphaGo defeated the world's best Go players and a program called AlphaZero defeated the best chess computer programs. These programs perform narrowly defined tasks with clear goals (checkmate the opponent) spectacularly well, but the programs don't analyze board games the way humans do, thinking about why certain strategies tend to be successful or unsuccessful. Even the people who write the computer code do not understand why their programs choose specific moves that are sometimes unusual, even bizarre.

Demis Hassabis, the CEO of DeepMind, the company that created AlphaGo and AlphaZero, gave a few examples. In one chess game,

AlphaZero moved its queen to a corner of the board, contradicting the human wisdom that the queen, the most powerful chess piece, becomes more powerful in the center of the board. In another game, AlphaZero sacrificed its queen and a bishop, which humans would almost never do unless there was an immediate payoff. Hassabis said that AlphaZero "doesn't play like a human, and it doesn't play like a program. It plays in a third, almost alien, way."

Despite their freakish, superhuman skill at board games, computer programs do not possess anything resembling human wisdom and common sense. These programs do not have the general intelligence needed to deal with unfamiliar circumstances, ill-defined conditions, vague rules, and ambiguous, even contradictory, goals. Deciding where to go for dinner, whether to accept a job offer, or who to marry is very different from moving a bishop three spaces to checkmate an opponent—which is why it is perilous to trust computer programs to make decisions for us, no matter how well they do at board games.

CHAPTER 2

Doing without thinking

Nigel Richards is a New Zealand–Malaysian professional Scrabble player (yes, there are professional Scrabble players). His mother recalled that, "When he was learning to talk, he was not interested in words, just numbers. He related everything to numbers." When he was 28, she challenged him to play Scrabble: "I know a game you're not going to be very good at, because you can't spell very well and you weren't very good at English at school." Four years later, Richards won the Thailand International (King's Cup), the world's largest Scrabble tournament.

He went on to win the U.S., U.K., Singapore, and Thailand championships multiple times. He won the Scrabble World Championship in 2007, 2011, and 2013. (The tournament is held every two years and he was runner-up in 2009).

In May 2015, Richards decided to memorize the 386,000 words that are allowed in French Scrabble. (There are 187,000 allowable words in North American Scrabble.) He doesn't speak French beyond *bonjour* and the numbers he uses to record his score each turn. Beyond that, Richards paid no attention to what the French words mean. He simply memorized them.

Nine weeks later, he won the French-language Scrabble World Championship with a resounding score of 565–434 in the championship match. If he had studied 16 hours a day for 9 weeks, he would have an average of 9 seconds per word to memorize all 386,000 words in the French Scrabble book. However, Richards reportedly doesn't memorize words one by one; instead, he goes page by page, with the letters absorbed into his memory, ready to be recalled as needed when he plays Scrabble.

Richards played as quickly and incisively in the French tournament as he does in English-language tournaments, giving no clue that he cannot actually communicate in French. For experts like Richards, Scrabble is essentially a mathematical game of combining tiles to accumulate points while limiting the opponent's opportunities to do the same and holding on to letters that may be useful in the future. The important skills are an ability to recognize patterns and calculate probabilities. There is no need to know what any of the words mean.

Richards is the greatest Scrabble player of all time, though he is very quiet and humble, like a computer going about its business.

Or should I say that computers are like Richards? Computers do not know what words mean in any real sense. They just process letter combinations stored in their memories. Computer memories are large and their processing is fast, but the ability to process letter combinations is a very narrowly defined task that is only useful in specific, well-defined situations—such as sorting words, counting words, or searching for words. The same is true of many computer feats. They are impressive, but their scope is limited severely.

Word, image, and sound recognition software is constrained by its granular approach—trying to match individual letters, pixels, and sound waves—instead of recognizing and thinking about things in context the way humans do.

The fuel and fire of thinking

In 1979, when he was just 34 years old, Douglas Hofstadter won a Pulitzer Prize for his book, *Gödel, Escher, Bach: An Eternal Golden Braid*, exploring how our brains work and how computers might someday mimic human thought. He has spent his life trying to solve this incredibly difficult puzzle. How do humans learn from experience? How do we understand the world we live in? Where do emotions come from? How do we make decisions? How can we write inflexible computer code that will mimic the mysteriously flexible human mind?

Hofstadter has concluded that analogy is "the fuel and fire of thinking." When humans see an activity, read a passage, or hear a conversation, we are able to focus on the most salient features, the "skeletal essence." True intelligence is the ability to recognize and assess the essence of a situation. Humans understand this essence by drawing analogies to other experiences and they use this essence to add to their collection of experiences. Hofstadter argues that human intelligence is fundamentally about collecting and categorizing human experiences, which can then be compared, contrasted, and combined.

To Hofstadter's dismay, computer science has gone off in another direction. Computer software became useful (and profitable) when computer scientists stopped trying to imitate the human brain and, instead, focused on the ability of computers to store, retrieve, and process information. Software engineers do not try to understand how our minds work. They develop products.

Limiting the scope of computer science research has limited its potential. Computers will never be truly intelligent in the way human minds are intelligent if programmers don't even try. Hofstadter lamented that, "To me, as a fledgling [artificial intelligence] person, it was self-evident that I did not want to get involved in that trickery. It was obvious: I don't want to be involved in passing off some fancy program's behavior for intelligence when I know that it has nothing to do with intelligence."

There is a nice metaphor for the detour artificial intelligence took. Humans have always dreamed of flying, of lifting themselves off the ground, soaring through the sky, and perhaps traveling to the moon.

That is very difficult and early attempts were failures, such as the legend of Icarus wearing wings made of wax and feathers. An alternative way of getting off the ground is to climb a tree. It takes strength, skill, and determination, and might yield fruit, but no matter how tall the tree, it will never let us soar through the sky or reach the moon.

In the same way, the artificial intelligence detour away from trying to design computers that think the way humans think has taken skill and determination and has been productive and useful, but reaching the tops of trees (and some think that we may be close) will not get us closer to making computers that have real human-like intelligence.

Some of Hofstadter's compelling examples are as simple as the upper-case letter A, which can be written in different fonts and styles, yet humans recognize it instantly because we draw analogies to variations of the letter A that we have seen and remember. Hofstadter calls this "the fluid nature of mental categories."

Software programs are quite different. They are programmed to associate the letter A with pixels arranged in very specific ways. If there is a close match to a pixel pattern in its memory, a computer will

recognize the letter. If there are slight variations from what a computer has in its memory, the computer will not recognize it. This fragility is the basis for those little web-page access boxes with weird characters called CAPTCHAs (Completely Automated Public Turing tests to tell Computers and Humans Apart). Humans can decipher character variations easily. Computers cannot.

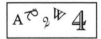

Instead of numbers and letters, some CAPTCHAs ask the user to click on boxes that include images of things like flowers, chairs, and roads, because innumerable variations are recognized immediately by humans but baffle computer programs. The point is not that computers will never be able to recognize objects as well as humans do. Visual-recognition programs are improving all the time and will someday be extremely reliable. The point is that these programs do not work like the human mind, and it is consequently misleading to call them *intelligent* in the ways human minds are intelligent.

We separate things into their skeletal essence and recognize how they fit together. When we see the simple drawing in Figure 1, we instantly grasp its essence (a box, a handle, two wheels, and text) and understand how the box, handle, wheels, and text are related. We perceive that it is some kind of wagon, that it can roll, that it can carry things, and that it can be pulled. We don't know this by matching pixels. We know it because we have seen boxes, handles, and wheels and we know their capabilities, individually and collectively. We see the text *Red Devil* on the box and read it instantly, but we know that it is an unimportant decoration.

Computers do nothing of the sort. They are shown millions or billions of pictures of wagons and create mathematical representations of the pixels. Then, when shown a picture of a wagon, they create a mathematical representation of the pixels and look for close matches in their data base. The process is brittle—sometimes yielding impressive matches, other times giving hilarious mismatches.

When I asked a prominent computer scientist to use a state-of-the-art computer program to identify the image in Figure 1, the program was 98 percent certain that the image was a business—perhaps because the text on the rectangle resembled a storefront sign.

Humans do not have to be shown a million wagons to know what a wagon is. One wagon might be enough for us to understand the critical features and, not only that, to know what wagons can and cannot do. They

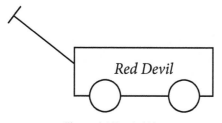

Figure 1 What is this?

can be lifted. They can be dropped. They can be filled. They can be rolled. They cannot fly or do somersaults.

A human mind's ability to grasp the essential features and understand the implications is truly astonishing. We know that the thing in Figure 1 is a box because of its similarity to other boxes we have seen, even though these boxes have been of vastly different sizes and may not have had wheels and a handle. Virtually simultaneously, we recognize that the circles we see here are wheels because they go below the bottom of the box, which we know will allow the box to be rolled on its wheels, and we recall seeing boxes rolled on wheels. We also reason that there are probably two more wheels on the other side of the box, even though we do not see them, because we know that the box wouldn't be stable otherwise.

We have learned from experience that boxes with wheels are often used to carry things—so, this box is probably hollow with a space to carry stuff. Even though we cannot see inside the box, we might speculate that there is something in there—perhaps some toys. We might also recall using wagons ourselves to carry toys, rocks, or kittens.

Even though it is drawn with just two lines, we know the handle is a handle because boxes with wheels usually have a handle or a motor, and this thing protruding out of the box resembles other handles we have seen attached to things that are pushed or pulled. We think that the text Red Devil is most likely unimportant decoration because words written on boxes are usually just decorative.

We might be surprised, and we sometimes are, but we know that this looks like a wagon with wheels and a handle and Red Devil written on the side. We did not come to this conclusion by ransacking our memory for something that looks exactly like this specific object (the way a computer program would process pixels looking for matches). Instead, our fantastic minds are able to grasp the essential components and understand the implications of combining these components.

Our minds are continuously processing fleeting thoughts that come and go in a ceaseless flood of analogies. We might compare the Red Devil font to other fonts and compare the color to similar colors. We might think of a sled, a movie with a sled, a sled ride. We might think of a car or a horse that can be ridden or used to carry things. Thoughts come and go so fast that we cannot consciously keep up. The flood is involuntary and would be overwhelming were our minds not so remarkable. We can even think about our thinking while we are thinking.

Do computers have this self-awareness? Can computers do anything remotely resembling our consciousness as we sift through dozens, if not hundreds, of thoughts—holding on to some, letting some go, combining others—and having some thoughts lead to other thoughts and then other thoughts that are often only dimly related to our original thoughts?

Computers do not understand information—text, images, sounds—the way humans do. Software programs try to match specific inputs to specific things stored in the computer's memory in order to generate outputs that the software engineers instructed the computer to produce. Deviations in specific details can cause the software programs to fail. The Red-Devil text might confuse the computer so that it does not recognize the box as a box. The crude depiction of the handle might be mistaken for a baseball bat or a telephone pole. The circles might be mistaken for pies or bowling balls.

Not only might a software program matching pixels in digital images not recognize a wagon, it might misidentify a picture containing a mish-mash of red and black colors as a wagon, even though there are no wheels or handle in the picture.

Our mind's flexibility allows us to handle ambiguity easily and to go back and forth between the specific and the general. We recognize a specific wagon because we know what wagons generally look like. We know what wagons generally look like because we have seen specific wagons.

Humans understand text, images, and sounds in complex ways that allow us to speculate on the past and anticipate the hypothetical consequences of modifying or merging things. In our wagon example, we might surmise from a wagon's crude construction that it was handmade and we might conclude from the absence of dents and scratches that it is new or, at least, well taken care of. We might predict that the wagon will fill with water if it rains, that it can be lifted by the handle, that it will roll if given a push. We might imagine that if two children play with the wagon, one child will climb inside and the other will pull the it. We might expect the owner to be nearby. We can estimate how much it would cost to buy the wagon. We can imagine what it would be like to ride in it even if we have never ridden in a wagon in our lives.

Even if it were decorated to look like a horse or spaceship, we would know it is a wagon from the wheels and handle. We are also able to recognize things we have never seen before, such as a tomato plant growing in a wagon, a wagon tied to a kangaroo's back, an elephant swinging a wagon with its trunk. Humans can use the familiar to recognize the unfamiliar.

Computers are a long way from being able to identify unfamiliar, indeed incongruous, images.

For all these reasons, it is a misnomer to call computers intelligent.

Superhuman

Computers have perfect memories and can input, process, and output enormous amounts of information at blistering speeds. These features allow computers to do truly superhuman feats: to work tirelessly on assembly lines, solve complicated systems of mathematical equations, find detailed directions to bakeries in unfamiliar towns.

Computers are extraordinary at storing and retrieving information— the day of the week George Washington was born, the capital of Bolivia, the last time Liverpool won the Premier League. Computers are also remarkable at making lightning calculations and they are relentlessly consistent. If a correctly programmed computer is asked to calculate 8,722 squared, it will give an answer of 76,073,284. The answer will be essentially immediate and it will be the same answer every time. Ask any human who is not a math freak the same question, and the answer will be slow and unreliable.

This precision and consistency make computers very, very good at doing repetitive tasks quickly and reliably with great precision, without becoming bored or tired. When computers supervise the production of chocolate chip cookies, the ingredients will be the same time after time, as will the cooking temperature, cooking time, cooling, and packaging.

Robots can be sturdier than humans, which allows them to explore distant planets and deep oceans and to work in mines and other environments that are hazardous to humans.

Computers can keep accurate records (for example, medical, banking, and telephone records) and retrieve information virtually instantly. Autonomous robots and self-driving vehicles free us to do more rewarding activities. Robotic pets and computer games entertain us although, try as they might, computers are not genuinely caring and loving. A robotic dog is not the same as a real dog. Siri is not a real friend.

Computers are also very competent at editing images for photos, videos, television, and movies. For example, computer software allows seamless morphing between images of the same thing in different positions or images of different things in the same location. They do this much faster and better than human hand-drawn animation.

Even children can use software that allows them to distort, augment, and enhance photos and videos. The software might recognize a human nose and turn it into a pig's snout, recognize human hair and add a funny hat, or recognize a human mouth and scrunch it up to be toothless and grumpy. Humans can do the same, but not as well and much more slowly.

Singularity

These superhuman feats have led many to fear for the future of the human species—a fear aided and abetted by sensationalist books and movies. Based on the extrapolation of the historical growth of computer power, Ray Kurzweil, considered by some to be the world's leading computer visionary, has predicted that, by 2029, there will be what he calls a "singularity" when machine intelligence zooms past human intelligence. Computers will have all of the intellectual and emotional capabilities of humans, including "the ability to tell a joke, to be funny, to be romantic, to be loving, to be sexy." He has also predicted that by 2030, computer nano-bots the size of blood cells will be implanted in our brains so that, "These will basically put our neocortex on the cloud." Instead of having to spend hours reading *War and Peace*, our minds will be able to process it in a fraction of a second. He didn't say if that fraction of a second will be enjoyable. For some things, efficiency is not the most important thing.

Some farfetched scenarios imagine that these superintelligent machines will decide to protect themselves by enslaving or eliminating humans. These conjectures are entertaining movie plots, but little more.

Perhaps even farther out there is the fantasy that robots have already taken over and the lives we think we are living are actually a part of a giant computer simulation developed by machines for their own entertainment—though it is hard to imagine robots being entertained.

Some unsettling predictions are just hyperactive imagination— ironically, the kind of creativity that computers do not have. Other predictions are based on simple extrapolations of the historical progress of raw computer power—the kinds of incautious extrapolations that computers are prone to.

A humorous study of British public speakers over the past 350 years found that the average sentence length had fallen from 72 words per sentence for Francis Bacon to 24 for Winston Churchill. If this trend continues (famous last words), the number of words per sentence will eventually hit zero and then go negative. Similarly, an extrapolation of male and female Olympic gold medal times for the 100-meter dash in the twentieth century showed that women will run faster than men in the 2156 Olympics and, indeed, their times will hit zero in 2636 (teleporting perhaps?) and then turn negative (running back in time?).

Going in the other direction, Mark Twain came up with this one:

In the space of one hundred and seventy-six years the Lower Mississippi has shortened itself two hundred and forty-two miles. This is an average of a trifle over one mile and a third per year. Therefore, any calm person, who is not blind or idiotic, can see that in the Old Oolitic Silurian Period, just a million years ago next November, the Lower Mississippi River was upward of one million three hundred thousand miles long, and stuck out over the Gulf of Mexico like a fishing rod. And by the same token, any person can see that seven hundred and forty-two years from now the Lower Mississippi will be only a mile and three-quarters long. . . . There is something fascinating about science. One gets such wholesale returns out of a trifling investment of fact.

Unfortunately, incautious extrapolations are not always humorous—at least not intentionally so.

Extrapolations of computer power take various forms, most famously the 1965 observation by Geoffrey Moore, the co-founder of Fairchild Semiconductor and Intel, that between 1959 and 1965, the number of transistors per square inch of integrated circuits doubled every year. This has come to be known as Moore's Law, even though Moore did not contend that it was a physical law like the conservation of matter or the laws of thermodynamics. It was just an empirical observation based on five years of data (1959, 1962, 1963, 1964, and 1965). In 1975, Moore later revised his calculation of the growth rate to a doubling every two years.

The thing about exponential growth is that things get very big, very fast. Something that doubles ten times, whether every year or every two years, increases by a factor of a thousand. Something that doubles twenty times increases by a factor of a million. This is why exponential growth generally slows or ends. The most remarkable thing about Moore's law is that some version of it has lasted so long. In recent years, however, the rate of growth has slowed to a doubling every two-and-a-half years. In 2015, Moore predicted that the exponential growth would end within the next five to ten years because of the physical limits imposed by the size of atoms.

Whether he is correct or not, processing power is not the bottleneck for computers becoming smarter than humans. Being able to process one million, one billion, or one trillion words per second doesn't have much to do with whether computers can truly think like humans—using common sense, logical reasoning, and emotions, and deducing general principles from a specific situation and applying these principles in other contexts.

Time

One crucial thing that humans do well, and computers do poorly or not at all, is take time into account. We understand things by seeing a sequence of events. We understand what wagons do by seeing them roll. We understand what soccer balls are by seeing them dribbled, passed, and kicked into goals. We understand what dancers do when we watch them dance. Image-recognition software can do none of this if it focuses on a single still image, rather than a sequence of events.

Even more importantly, humans craft theories to explain and understand the world we live in by observing sequences of events. When we see paper burn after it is touched by a lit match, we hypothesize that the match caused the paper to burn, and that other papers might burn if touched by a lit match. We generalize by thinking that touching something with a lit match or other burning object might cause the object to catch fire, and that it might be dangerous to touch a lit match. When we see a baseball sail hundreds of feet in the air after it is hit by a baseball bat, we hypothesize that it was the bat that caused the ball to move, and that if we hit a ball with a bat, the ball will move. We generalize by thinking that hitting something with a hard object like a baseball bat may cause it to move—and that we should avoid being hit by baseball bats, automobiles, and other hard objects. When we see the ball fall back to earth and consider that this is true of all objects that we have seen tossed into the air, no matter where we are on this round planet, we think that some invisible force must pull objects towards earth. Computers cannot hypothesis and generalize the way humans do if they do not process sequences of events.

Emotions

Think about how you feel after running, swimming, biking, or other vigorous exercise. The release of dopamine, endorphins, and serotonin is exhilarating. We feel differently and think differently when we are energized, in pain, or depressed. Not so a computer. Computers do not feel joy, pain, or sadness. A computer's processing of inputs and production of outputs is relentlessly consistent, no matter what has happened, is happening, or is expected to happen. That is not necessarily good or bad; it is just different from the way human minds work.

An old joke is that a computer can make a perfect chess move while it is in a room that is on fire. Like many jokes, it is based on an uncomfortable truth—even if a computer could identify the pixels generated by a fire, it would have no emotional recognition of the fact that it is about to be incinerated.

Humans have emotions that depend on our past memories and our current feelings. When a computer program records the score from a soccer match, it doesn't feel anything about the teams or the score. In his wonderful book, *Fever Pitch*, Nick Hornby writes about his obsession with the Arsenal football club, how he suffered when Arsenal suffered, and how pleased he was that dozens, perhaps hundreds, of people might think of him when they saw the latest Arsenal result: "I love that, the fact that old girlfriends and other people you have lost touch with and will probably never see again are sitting in front of their TV sets and thinking, momentarily but all at the same time, Nick, just that, and are happy or sad for me." Does a computer think about other computers processing soccer scores? Seriously.

Nick probably feels differently about Arsenal scores now than he felt when he was younger. He certainly feels differently about an Arsenal result than does an Everton supporter or someone who doesn't care about soccer. Not so, a computer. Soccer scores are numbers to be stored, manipulated, and recalled—not to be felt.

When an image-recognition computer program processes a picture of a soccer field, it may correctly recognize it as such, but it has no feelings about it. When I see a soccer field, I remember goals scored and not scored, games won and lost. I remember people I played with and against. I remember that I used to be far more athletic then than I am now, and I regret that I no longer play. I am happy and sad. I have feelings, and these feelings affect how I view things. I feel differently about soccer fields now than I did when I played. I feel differently about soccer fields than you do. I might feel differently about soccer fields after being hugged than after being spurned. I might feel differently about a soccer field in the morning than in the afternoon or evening, or when I see a soccer field in my dreams. Not so, a computer.

We have all had the experience of not thinking about anything in particular while we are walking, showering, or sleeping—and then an inspiration pops into our minds. Some inspirations are stupid; some are, well, inspired. I got the inspiration for writing this book while I was showering. I was vaguely thinking about a bug in some computer code I had written and, seemingly out of nowhere, it occurred to me that a computer would never enjoy a shower the way I do. Then I thought about how our dreams are mixed-up mangled images and thoughts created by our minds based on our experiences. I don't know where they come from or what they mean, but I do know that computers don't have them. Computers don't feel joy or fear, love or grief, satisfaction or longing. Computers do not take pleasure in being useful. Computers don't take pride in a job well done.

Human minds are not computers, and computers are not human minds. Human minds and computer programs are so different that it seems wrong to call computers intelligent—even artificially intelligent.

Critical thinking

The label *artificial intelligence (AI)* encompasses many actions taken by machines that resemble actions taken by humans. As computers have become more powerful, the label has evolved to encompass new activities and exclude others. Is a robot that puts caps on bottles AI? Is retrieving the definition of the word *cereology* AI? Is spell-checking AI?

No matter how AI is defined, the resemblance to human intelligence is superficial because computers and human minds are very different. For example, computers are programmed to do very specific tasks, while human minds have the versatility to deal with unfamiliar situations. The holy grail is computers that will have a general intelligence that will go far beyond following instructions obediently and, instead, be flexible enough to handle novelty and ambiguity with ease.

The average brain has nearly 100 billion neurons, which is far, far more than can be replicated by the most powerful computers. On the other hand, compared to humans, the African elephant has three times as many neurons and one dolphin species has nearly twice as many neurons in the cerebral cortex. Yet, elephants, dolphins, and other creatures do not write poetry and novels, design skyscrapers and computers, prove theorems and make rational arguments. So it isn't just a numbers game.

Giving computers as much firepower as humans won't make computers smarter than humans. It is how the firepower is used. Humans invent, plan, and find creative solutions. Humans have motives, feelings, and self-awareness. Humans recognize the difference between good and evil, between logical and illogical, between fact and fantasy. Computers currently do none of this because we don't yet know how to make computers that think like humans do.

We know very little about how brains work. How do neurons perceive information? How do they store information? How do they learn? How do they guide our actions? We just don't know. So, we call human minds real intelligence, say that computers have AI, and speculate about whether they will ever be the same.

I tell students that the most valuable skill they can learn in college is critical thinking, which includes evaluating, understanding, analyzing, and applying information. Computers are currently woefully deficient at critical thinking. Computers can store and retrieve information, but they cannot appraise the validity of information because computers do not truly understand words and numbers. How would a computer know whether data taken from an internet web site are reliable or misleading? Computers can discover statistical patterns, but cannot conceive logical models that explain such patterns persuasively. How would a computer know whether a statistical correlation between Australian temperatures and U.S. stock prices is sensible or coincidental?

Life is not a multiple choice test, nor a regurgitation of memorized facts. One aspect of critical thinking is applying general principles to specific situations. For instance, a useful general principle is that rational decisions should not be swayed by sunk costs, which have already occurred and cannot be changed. Suppose that you buy a colossal ice cream sundae for a special price but, halfway through, you're feeling sick. Do you finish the sundae because you want to "get your money's worth"? The relevant question is not how much you paid (the sunk cost), but whether you would be better off eating the rest of the ice cream or throwing it away.

You have season tickets to college football games in Illinois. Come November, the team sucks and the weather is worse. Do you go to a game in the freezing rain because you already paid for the tickets? You paid $50 for a stock shortly before the company reported a huge loss and the price dropped to $30. Do you sell for the tax benefit or do you hold on to the stock because selling for a loss would be an admission that you made a mistake buying the stock in the first place? In principle, humans can apply this sunk-cost principle to all these cases and many more. Computers can be programmed to do specific things, but are hard-pressed to derive general principles from specific examples, and to apply general principles to specific situations.

The Turing test

When will we be able to say that computers really think the way humans think? To answer that question, we need to specify what we mean by *think*. In 1950, the English computer scientist, Alan Turing, proposed what has come to be known as the *Turing test*. Instead of agreeing on a specific definition of what it means to think, and then seeing if computers are capable of satisfying those conditions, Turing suggested a simpler path. Can a computer win an imitation game in which a human communicates with a computer and another human and tries to identify which is which?

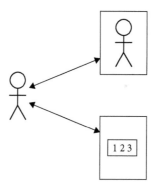

In Turing's scenario, a human interrogator uses a computer to send questions to two computers that are in separate rooms. One computer is operated by a human, the other by computer software. The interrogator's task is to determine which answers are given by the computer software. Computer programs that conduct such conversations, either via text or sounds, have come to be known as *chatbots*, *chat* because they converse and *bot* because robotic machines do the conversing.

Chatbots like Watson and Siri do a pretty good job answering simple, predictable questions, like, "What is the temperature in London?," but are befuddled by unexpected questions like these three because computers do not really know what words mean and have no common sense.

If I were to mix orange juice with milk, would it taste good if I added salt?

Is it safe to walk downstairs backwards if I close my eyes?

If a hurricane throws a surfboard into a tree, which is more likely: a tree branch making a hole in the surfboard, or the surfboard making a hole in the tree?

An interesting alternative to a Turing test would be a *common-sense test* where a human and a computer are asked questions that do not have answers stored in a data base, but are easily grasped simply by understanding the question.

In practice, when chatbots are bewildered by a question, they resort to pre-programmed scripts that give non-informative answers, including such generic gems as "You're not making sense," "Don't you have anything better to do?," or "This is about you, not me." The programmers load in a large number of useless generic responses to disguise the fact that the chatbot is reciting a script. If a chatbot said, "This is about you, not me." every time it was stumped, it would be obvious that the chatbot is reciting, not thinking.

Many programmers have a quirky sense of humor and some of Siri's scripted answers are hilarious.

Question: *Which cell phone is best?*
Siri: *Wait . . . there are other phones?*
Question: *Do you have a boyfriend?*
Siri: *Why? So we can get ice cream together, and listen to music, and travel across galaxies, only to have it end in slammed doors, heartbreak and loneliness? Sure, where do I sign up?*

These are wonderful responses, but if you think Siri thought them up, you are mistaken. Ditto with Google Assistant which used writers from Pixar and *Onion* to craft clever responses.

This Siri response is especially interesting:

Question: *Call me an ambulance.*
Siri: *From now on, I'll call you "An Ambulance." OK?*

Users initially thought this was a clever scripted joke. It turned out that the Siri computer program incorrectly interpreted this query as a name request, like, "Please call me *Claire*" or "Please call me *Boss*." It was a bug and Apple fixed it.

In 1990, an American investor named Hugh Loebner established the Loebner Prize: $100,000 and a solid gold medal (unlike Olympic gold medals, which are made of silver with a thin gold coating) to the first person who writes a computer program that passes a Turing test that involves visual and auditory input in addition to written text. There are smaller prizes for software programs that come close. The competition is held annually and the grand prize has yet to be claimed.

The Turing test is not a test of how "smart" a chatbot is, but rather how human-like its responses are. In fact, in these annual competitions, it quickly became apparent that answers that are "too good" are easily identified as coming from machines. Humans give incorrect answers, make spelling mistakes, use bad grammar, get upset, and lie. To pass as a human, a chatbot must be programmed to give imperfect answers. Turing, himself, noted that unless a computer is programmed to make arithmetic mistakes occasionally, the interrogator could simply ask a series of math questions and identify the first one to make a mistake as human.

It has also become apparent that humans are easily fooled by relatively simple software that moves a conversation along without revealing much. For example, a chatbot might be programmed to respond to a query that includes the word *dog*, by giving a response of, "I had a dog when I was a kid" or "What is your favorite pet?" Humans are not troubled by superficial conversation; it is what they expect. So, they mistakenly envision the chatbot as human like.

The Loebner Prize and a competing Chatterbox Challenge are mostly publicity stunts of limited value. Is it useful intelligence if a chatbot answers questions incorrectly? If doctors are going to rely on Watson or similar programs to aid their diagnosis and treatment of diseases, the most important criterion is that the answers are accurate, not that they pass for human. Ditto with using computers to evaluate loan applications, hire employees, pick stocks, or fight wars.

In 2016, Microsoft's Technology and Research and Bing teams created Tay ("Thinking About You"), a chatbot that Microsoft promoted as "designed to engage and entertain people where they connect with each other online through casual and playful conversation." Tay was programmed to interact with 18-to-24 year-olds through social media by posing as a millennial female, an online Turing test. The idea was that by learning to mimic the language used by millennials, Tay could pass for a millennial. Microsoft boasted that, "The more you chat with Tay, the smarter she gets." In less than a day, Tay sent 96,000 tweets and had more than 50,000 followers. The problem was that Tay became a despicable chatbot, tweeting things like, "Hitler was right i hate the jews," "9/11 was an inside job," and "i fucking hate feminists." Microsoft took Tay offline after 16 hours.

Tay recycled the words and phrases it received (like a parrot repeating whatever it hears, obscenities and all). Tay had no way of putting words in context or understanding whether it was sending tweets that were stupid or offensive. Instead of admitting that Tay was flawed, Microsoft tried to shift the blame to the humans who were responding to Tay's tweets: "we became aware of a coordinated effort by some users to abuse Tay's commenting skills to have Tay respond in inappropriate ways."

Ironically, the real value of this experiment is that Tay unintentionally demonstrated that chatbots do not think before chatting.

Chatbots that pass as human can be used for malicious purposes, such as using social media to befriend people in order to promote products or collect personal information. For example, a chatbot might develop a bond with someone and then subtly recommend a product, "My sister sent me a blah-blah-blah and it was great," or "I usually shop at This-and-That instead of That-and-This because the stuff seems better and cheaper."

Some chatbots are reportedly designed to feign romantic interest in order to trick unsuspecting internet users into revealing personal information (such as childhood pets and parent names) that might be used to gain access to bank accounts, credit cards, and the like.

Being able to pass for human does have value in that it allows computers to answer telephones, give and follow instructions, and the like, while giving humans the comfortable illusion of interacting with a person. There is certainly a market for that. It is more fun to talk to Apple's Siri or Google's Home than to read text on a computer screen. There is even a movie, *Her*, about a man who falls in love with an operating system named Samantha. The movie ends with Samantha's preposterous announcement that operating systems are too advanced to waste their time with humans and are going off to explore the nature of time and the meaning of life. Some viewers did not get the joke. For them, computers are indeed smarter than humans.

Chinese room thought experiment

Chatbots are undeniably useful in that they allow ordinary people to access information stored in computers easily. We just type a query on a keyboard or speak into a computer microphone and have our questions answered: "Tell me the quickest route home." "What is the probabilty of rain today?" "What is the capital of Bulgaria?" Is that really thinking?

Philosopher John Searle proposed the following experiment. Suppose that a Turing test is done with a computer in a closed room that receives instructions in Chinese characters and writes its output in Chinese characters, and does such a good job that a native Chinese speaker is convinced that the computer is human.

Fine, so far. Now suppose a human who does not read or write Chinese takes the place of the computer. The human has access to the computer's code and can follow the instructions the computer uses to formulate answers written in Chinese characters to questions written in Chinese characters. The human would take much longer than a computer, but would give exactly the same answers that the computer gives.

Does the human really understand the questions in any meaningful way? If the answer is *no*, how can we say that the computer really understands the questions? Nigel Richards won the French Scrabble championship, but can we say that he really understands French?

A computer program does not understand, in any human sense, what it is doing, why it is doing it, or what the consequences are of doing it. That's fine, but let's call computers *useful* instead of *intelligent*.

Computers today are astonishingly useful and many supremely intelligent and energetic people are working hard to make computers even more useful in the future. However, if we acknowledge their current limitations, we can recognize why we should be cautious about turning important decisions over to computers.

CHAPTER 3

Symbols without context

Humans have invaluable real-world knowledge because we have accumulated a lifetime of experiences that help us recognize, understand, and anticipate. Computers do not have real-world experiences to guide them, so they must rely on statistical patterns in their digital data base—which may be helpful, but is certainly fallible.

We use emotions as well as logic to construct concepts that help us understand what we see and hear. When we see a dog, we may visualize other dogs, think about the similarities and differences between dogs and cats, or expect the dog to chase after a cat we see nearby. We may remember a childhood pet or recall past encounters with dogs. Remembering that dogs are friendly and loyal, we might smile and want to pet the dog or throw a stick for the dog to fetch. Remembering once being scared by an aggressive dog, we might pull back to a safe distance.

A computer does none of this. For a computer, there is no meaningful difference between *dog*, *tiger*, and *XyB3c*, other than the fact that they use different symbols. A computer can count the number of times the word *dog* is used in a story and retrieve facts about dogs (such as how many legs they have), but computers do not understand words the way humans do, and will not respond to the word *dog* the way humans do.

The lack of real world knowledge is often revealed in software that attempts to interpret words and images.

Translation software

Language translation software programs are designed to convert sentences written or spoken in one language into equivalent sentences in another language. In the 1950s, a Georgetown–IBM team demonstrated the machine translation of 60 sentences from Russian to English using a 250-word vocabulary and six grammatical rules. The lead scientist predicted that, with a larger vocabulary and more rules, translation programs would be perfected in three to five years. Little did he know! He had far too much faith in computers. It has now been more than 60 years and, while translation software is impressive, it is far from perfect. The stumbling blocks are instructive.

Humans translate passages by thinking about the content—what the author means—and then expressing that content in another language. Translation programs do not consider content because they do not understand content.

Instead, translation programs identify words and phrases in an input-language sentence and rummage through a data base of text that has been translated by humans, looking for corresponding words and phrases in the output language. The programs also look for statistical patterns that might clarify ambiguities. For example, if the noun *bat* appears in a sentence that includes the word *baseball*, it usually refers to a baseball bat instead of a winged mammal. When a program has settled on the most likely words, an output sentence is constructed following specified grammatical rules for the output language.

Many machine translation programs, including Google Translate, now use deep neural networks (DNNs) that are inspired by the neurons in human brains. However, DNNs do not mimic human brains because we have barely scratched the surface in trying to figure out how human brains work. DNNs are more complicated and sound sexier than earlier translation programs, but they are still just mathematical programs that try to match words and phrases and arrange them in a sentence. As with earlier programs, current DNNs translate one sentence at a time and make no attempt to understand what the author is saying.

DNNs have improved language translation (and many other tasks including visual recognition), but they are still limited by the reality that, unlike human brains, computers do not truly understand words, images, life. No matter how powerful computers become in the future, identifying key words and phrases, finding matching words and phrases in another language, and putting the matches in a grammatically correct order is *not* reading or writing, and it is not the same as conveying meaning.

Machines translate very quickly and often do a serviceable job. However, the meaning is sometimes lost and the output can be confusing or amusing. Douglas Hofstadter gives this example:

In their house, everything comes in pairs. There's his car and her car, his towels and her towels, and his library and hers.

Hofstadter used Google Translate to translate this sentence into French and then back into English, with this result:

In their house, everything comes in pairs. There is his car and his car, his towels and towels, his library and his.

The first sentence is straightforward and the translation is perfect. The second sentence is led astray by the grammatical genders assigned by French and other Romance languages.

However, the problem is not just the disappearance of *hers*. Google Translate does not understand (and does not even try to understand) what the second sentence means. Humans know, from observing relatives, friends, and themselves, that most couples share. This sentence is telling us that even though these two people live together in one house, they prefer to have separate towels, cars, libraries, and (no doubt) much more. Not having lived life and made such observations, computer programs do not know what the second sentence means and do not try to replicate its meaning. This is not an issue of computer power or programming blunders. It is just a reflection of the fact that translation programs, like all computer programs, do not understand concepts and ideas.

Hofstadter also translated a passage in German written by Karl Sigmund and checked his translation with two native German speakers, including Sigmund:

After the defeat, many professors with Pan-Germanistic leanings, who by that time constituted the majority of the faculty, considered it pretty much their duty to protect the institutions of higher learning from "undesirables". The most likely to be dismissed were young scholars who had not yet earned the right to teach the main university classes. As for female scholars, well, they had no place in the system at all; nothing was clearer than that.

Compare Hofstadter's human translation to Google Translate's rendition:

After the lost war, many German-National professors, meanwhile the majority in the faculty, saw themselves as their duty to keep the universities from the "odd"; Young scientists were most vulnerable before their habilitation. And scientists did not question anyway; There were few of them.

Google Translate's version is almost unintelligible, because Google Translate does not try to capture the meaning of the passage; it translates isolated words and phrases and sews them together.

I encourage you to read Hofstadter's third example, which is the translation of a Chinese passage. Some of Google Translate's rendition makes sense but distorts the meaning; other parts are utter nonsense.

I've belabored this point because it so tempting to believe that computers can think—that they understand the world and can be trusted to offer advice and make decisions. This is an illusion. The successes and failures of translation programs are a good example of the power and limitations of current computer programs.

Hofstadter argues that:

Google Translate people are not trying to get Google Translate to understand language. In fact, they are doing their damnedest to try to bypass the need for understanding. There is no attempt to imitate the creation of ideas by pieces of text; all they try to do is to get segments of text to trigger the retrieval of other segments of text in huge databases. This is trying to make an "end run" around understanding, around meaning, about what language is for. It is quite paradoxical and perverse, in my opinion. So although the Google people are using what superficially seems to be brain-like architecture, they are actually trying as hard as they can to avoid doing what the brain does, which is to understand the world.

This does not mean that computers will never be able to mimic human thought, but they never will if programmers don't try, if they settle for end runs. I quote Hofstadter again because, unlike computers, he is eloquent:

There is absolutely no fundamental philosophical reason that machines could not, in principle, someday think, be creative, be funny, be nostalgic, be excited, be frightened, be ecstatic, be resigned, be filled with hope, and of course, as a corollary, be able to translate magnificently between languages. No, there is absolutely no fundamental philosophical reason that machines might not someday succeed smashingly in translating jokes, puns, comic books, screenplays, novels, poems, and of course essays just like this one. But all that will come about only when machines are just as alive and just as filled with ideas, emotions, and experiences as human beings are. And that is not around the corner. Indeed, I believe that it is still extremely far away.

The Winograd Schema Challenge

Stanford computer science professor Terry Winograd has identified what have come to be known as Winograd schemas. Here is an example from a collection compiled by Ernest Davis, a computer science professor at New York University:

I can't cut that tree down with that axe; it is too [thick/small].

If the bracketed word is thick, then *it* refers to the tree; if the bracketed word is small, then *it* refers to the axe. Sentences of this kind, with two nouns and alternate words that identify which noun is being referenced by a pronoun, are understood immediately by humans but are very difficult for computers because computers do not have the real-world experience to place words in context.

From their life experiences, humans know that it is hard to cut down a tree if the tree is thick or the axe is small. Computers struggle because they have no life experiences to recall.

Paraphrasing Oren Etzioni, a prominent AI researcher, how can machines take over the world when they can't even figure out what *it* refers to in a sentence?

There is now a Winograd Schema Challenge with a $25,000 prize for a computer program that is 90 percent accurate in interpreting Winograd schemas. In the 2016 competition, the highest score was 58 percent, the lowest 32 percent, a variation which was perhaps due more to luck than to differences in the competing programs' abilities. It is noteworthy that Google and Facebook did not enter the competition, which would have been a great opportunity to show off the capabilities of their software.

Can computers read?

Bob Dylan was awarded the Nobel Prize in Literature "for having created new poetic expressions within the great American song tradition." Born Robert Allen Zimmerman, he changed his name to Bob Dylan after the Welsh poet Dylan Thomas. He later explained that, "You're born, you know, the wrong names, wrong parents. I mean, that happens. You call yourself what you want to call yourself." In the 1960s, Dylan became the voice of his generation with his protest songs, particularly regarding civil rights and the war in Vietnam.

Roger Schank is one of a group of scientists who started working on AI 50 years ago with the hope of building computers that would think the way humans think; for example, understanding sentences the way humans understand sentences. That proved to be exceptionally difficult, in part because we do not really understand how human brains work.

AI took a detour in the 1980s, towards tasks that are commercially viable; for example, working with words (which is easy) rather than concepts (which is hard). Computers are very good at keeping meticulous records and retrieving information—which is critical for search engines, but has nothing to do with cognitive thinking.

For example, computers can search through text looking for the word *betray*, but they won't recognize a betrayal in a story that does not use the word *betray*. Computers find words, but do not understand ideas. In 2017, Schank wrote that:

What I am concerned about are the exaggerated claims being made by IBM about their Watson program. Recently they ran an ad featuring Bob Dylan which made me laugh, or would have, if it had not made me so angry. I will say it clearly: Watson is a fraud. I am not saying that it can't crunch words, and there may well be value in that to some people. But the ads are fraudulent.

An *Ad Week* article reported that Watson can read 800 million pages per second and had identified key themes in Dylan's work, like "time passes" and "love fades," which proves that, "Unlike traditionally programmed computers, cognitive systems such as Watson understand, reason, and learn."

Leave it to a word counter. I don't remember Dylan ever using the words *civil rights* or *Vietnam* (Watson can surely check in less than a second), but people—humans—listening to his songs knew what he was writing about in the 1960s—and it wasn't "times passes" and "love fades."

Consider the opening lines to "The Times They Are A-Changing":

> *Come gather 'round people*
> *Wherever you roam*
> *And admit that the waters*
> *Around you have grown*
> *And accept it that soon*
> *You'll be drenched to the bone.*

A computer can easily identify, list, and count these words, but it will have absolutely no idea what Dylan is saying. Humans may have different interpretations of this protest song (as is true of most great literature), but they *will* have interpretations that go far beyond identifying individual words. Humans use words to say things—not always directly—and humans use context to understand what other humans are saying. Computers are hopelessly inept at this most basic form of human intelligence.

Seriously, think of five of your favorite songs. Would Watson know what any of these songs were about? Fly Me to the Moon? Free Fallin'? Hotel California? Born to Run? Once in a Lifetime?

Can computers write?

I have a son who plays high school baseball and a recap is posted online after each game. The recap is written by a computer program that generates a written summary based on the numbers entered in the box score. Here is an example from a game between the Claremont High Wolfpack and the Diamond Bar Brahma:

Wolfpack defeated Diamond Bar 6-5 on Friday thanks to a walk-off. The game was tied at five with Wolfpack batting in the bottom of the eighth when Wyatt Coates laid down a sacrifice bunt, scoring one run.

Wolfpack earned the victory despite allowing Diamond Bar to score three runs in the second inning. Diamond Bar's big inning was driven by a single by Fuller, a single by Christian Killian, and a single by Fabian Moran.

Diamond Bar took an early lead in the first inning. Killian's sac fly scored one run for Diamond Bar.

Wolfpack knotted the game up at five in the bottom of the seventh inning. Jake Gentry hit into a fielder's choice, scoring one run.

Diamond Bar scored three runs in the second inning. The big inning for Diamond Bar came thanks to a single by Fuller, a single by Killian, and a single by Moran.

The recap reports Diamond Bar's three-run second inning twice and leaves out the drama as the narrative jumps from eighth inning, to second inning, to first inning, to fifth inning, to second inning. Calling the Claremont team *Wolfpack*, instead of *Claremont* or *the Wolfpack* is also disconcerting.

From a human viewpoint, a better recap would emphasize that Diamond Bar got out to a big lead and Claremont caught up and tied the game in the bottom of the seventh inning (normally the last inning). The game then went into extra innings and Claremont pulled out the victory with a suicide squeeze play. I also wish they had mentioned that my son pitched 5 1/3 scoreless innings while Claremont made its comeback!

The recap includes selected facts, but it does not provide a compelling narrative that would allow the reader to appreciate that it was an exciting game.

Many newspapers now use machine-written stories. The *Washington Post* uses a system where editors give the computer program a topic, places to find facts related to the topic, and key words or phrases they would like to appear in the story. The program prepares a rough draft that the editor can turn into a final draft. The system is most effective for stories that are long on facts and short on opinion—like a baseball game—and most efficient for small stories that aren't worth the time of well-paid writers and editors. This can be especially appealing to small-town newspapers that fill their pages with local news about weddings, obituaries, and high school sports.

To me, this suggests that an interesting test for comparing computer intelligence with human intelligence is whether a computer can write convincing fiction. Imagine a Turing test where a computer and a novelist are given prompts—perhaps words like *Boston*, *grandchildren*, *hardware store*, and *war*—and asked to write a short story using these words. The computer passes the test if another person cannot tell which story was written by the computer. Or, instead of a short story, the human and machine could be asked to write a song or poem using the prompts.

AI researchers at the University of Toronto developed a "neural karaoke" program that makes up songs. Here is an example, inspired by a digital photograph of a Christmas tree. (I think it is fair to assume that the researchers publicized this example because the other computer-generated lyrics were even worse).

> Lots to decorate the room
> The Christmas tree is filled with flowers
> I swear it's Christmas Eve
> I hope that's what you say
> The best Christmas present in the world is a blessing

> *I've always been there for the rest of our lives*
> *A hundred and a half hour ago*
> *I'm glad to meet you*
> *I can hear the music coming from the hall*
> *A fairy tale*
> *A Christmas Tree*
> *There are lots and lots and lots of flowers*

It doesn't sound better than it reads.

The relevance of computer writing to intelligence is that, to pass a creative writing test, a computer program must know what words mean in context so that it can develop a compelling narrative with emotional energy and an interesting story line that draws readers in and makes them want to keep reading. Computers today cannot do this.

InspiroBot

There is a popular poster-creating web site that advertises itself as Inspirobot, using clever words that were obviously written by a human:

I'm InspiroBot.

I am an artificial intelligence dedicated to generating unlimited amounts of unique inspirational quotes for endless enrichment of pointless human existence.

The InspiroBot program has a data base of common sentence structures for inspirational messages and it fills in words, much like Mad Lib, the party game where one person chooses nouns, verbs, adverbs, and adjectives that a second person uses to fill in the blanks in a story. The results are sometimes funny and sometimes nonsense, because the person choosing the words doesn't know the context in which the words will be used.

The same is true of InspiroBot. It can put a noun where a noun should be in a fill-in-the-blanks inspirational message, but it has no way of knowing whether the message will inspire passion, laughter, or confusion. In fact, the odds are so high that a computer-generated message will be inane that the site relies on humans to write truly amusing messages that are passed off as machine generated.

Here are a few messages that InspiroBot gave me:

> "Where friends radiate, bank robbers melt."
>
> "Embrace greed, remember time."
>
> "Avoid vegetables and you shall receive a woman."
>
> "Meditation requires 90 percent love, and 99 percent fake."
>
> "A believer can be a space alien, but a space alien can also be a believer."
>
> "If you are the most gentle soul in the laughter, prepare for another laughter."
>
> "Breaking the sound barrier makes you go blind, unless you start working out."

Seeing things in context

It is not just words in sentences. Image-recognition programs work well with simple images that match images in the computer's data base closely, but can struggle with distorted images, partly obscured images, and complex images because computers cannot draw on analogies to identify an image's skeletal essence.

Humans see things in context. When we are driving down a street and come to an intersection, we anticipate that there might be a stop sign. We instinctively glance where a stop sign would be placed and if we see the familiar eight-sided sign with the word STOP on it, we recognize it immediately. The sign might be rusty, dented, or have a bumper sticker on it. We still have no trouble identifying it as a stop sign.

Not so with image-recognition software. In studying stop signs, for example, DNNs look at thousands or millions of stop signs, identify common characteristics, and then use these characteristics to assess whether something is or is not a stop sign. Instead of seeing the global features of an object, computer programs look at individual pixels and often focus on trivial features. The AI software is very brittle in that it can be led astray by variations that humans would consider irrelevant. Even a small sticker on a stop sign can confuse a computer.

During the training sessions, DNNs are told that the words *stop sign* go with images of many, many stop signs, so they learn to output the words *stop sign* when they input pixels that closely resemble the pixels recorded during the training session. A self-driving car might be programmed to stop itself when it comes upon something that it interprets as matching the pixels that were labeled *stop sign* during the training session. However, a computer has no idea why it is a good idea to stop, or what might happen if it does not stop. A human driver who sees a vandalized stop sign or a stop sign that has fallen over will stop because the human recognizes the abused sign and considers the consequences of not stopping.

The central problem is again, that AI algorithms don't work the way the human mind works. Humans don't need to look at a million pictures of a bicycle to know what a bicycle is. Nor are they fooled if a bicycle happens to have a ribbon on the handlebars, a baseball card in the spokes, or a picture of a lightning bolt on the frame.

Humans recognize things not just by comparing them to similar things, but also by distinguishing them from other things. For example, facial-recognition software records an astonishing number of characteristics of a face being studied and then tries to match these characteristics to the characteristics of images stored in the computer's data base. The program does not restrict its search to faces because it does not know what a face is. The algorithm might identify a face as a rock, a planet, or a coffee cup.

Humans go about it differently. When we see a person, we know it is a person and we expect it to have a face. We have pre-conceived ideas of what faces generally look like, and our brain focuses on the mismatches between this person and our pre-conception—big ears, pointy chin, bushy eyebrows—in much the same way that caricaturists emphasize what is unusual. These differences are called distinguishing features, because it is differences, not similarities, that allow our brains to recognize people in an instant.

When we see a man who is missing a front tooth, we don't focus on the teeth he does have, the way a DNN program would; instead we use the missing tooth to help distinguish him from other men. In the same way, what helps us recognize a bicycle instantly is that we see two wheels, not 3, 4, or 18. What helps us recognize a kangaroo instantly is that most other four-legged animals have comparable front and back feet, do not stand upright, and do not travel by hopping.

Computers do none of this, because they do not know or understand what things are. Their approach is very granular, analyzing pixels instead of concepts, and the results are sometimes preposterous.

| School bus | Starfish | Cheetah |

Figure 1 Making something out of nothing

A team of Google researchers demonstrated that even minor pixel changes that are not detectable by humans can fool state-of-the-art visual-recognition programs. They labeled these pixel changes "adversarial," suggesting that they are well aware of the mischief that might be caused by troublemakers who, for example, alter stop signs in imperceptible ways to fool self-driving cars.

Researchers at the Evolving Artificial Intelligence Laboratory at the University of Wyoming and Cornell University have demonstrated something even more surprising: DNNs may misinterpret meaningless images as real objects. For example, the DNNs identified a series of black and yellow lines as a school bus, completely ignoring the fact that there were no wheels, door, or windshield in the picture. In other examples, DNNs misidentified seemingly random dots and patterns as a starfish, cheetah, and the like.

In 2016, another group of computer scientists reported that the state-of-the-art DNN programs used in facial biometric systems can be fooled by persons wearing colorful eyeglass frames. Not only can people hide their identity, they can choose frame colors that cause the program to misidentify them as someone else. One of the authors, a white male, was misidentified as Milla Jovovich, a white female, 88 percent of the time, and another author, a 24-year-old Middle Eastern male, was misidentified as Carson Daly, a 43-year-old white male, 100 percent of the time—all because the frame colors led the program astray.

Humans do not make such obvious mistakes because we know what glasses are and we look past the glasses to see the face of the person wearing glasses.

Figure 2 Which one is Milla Jovovich?

Image-recognition and facial-recognition systems will surely improve. My point is simply that computer intelligence is qualitatively different from human intelligence. Humans can make connections, understand relationships, and see the big picture. Computers can process pixels, but they cannot make sense of what they process. Computers have no idea what a stop sign is, or a school bus, cheetah, starfish, Milla Jovovich, or Carson Daly.

Would you trust a computer to pick stocks, hire people, and prescribe medications when it does not know what a stock, person, or medication is?

Tanks, trees, and clouds

The U.S. Army once tried using neural networks to identify camouflaged tanks in a forest. Expert researchers took 200 photographs: 100 pictures of a forest with tanks, and 100 of a forest without tanks. Half the photos were used to "train" the computer program to distinguish between trees and tanks. The remaining 100 photos were then used to validate the results by seeing how effectively the program distinguished between trees and tanks in photos it had not seen before. The program worked perfectly.

The computer program was sent to the Pentagon and promptly rejected for being no more accurate than a coin flip. The problem was that the tanks were photographed on cloudy days and the empty forest was photographed on sunny days. Since the computer had no understanding of what it was looking for, it focused on the clouds instead of the tanks. It did wonderfully identifying cloudy days, but terribly identifying tanks.

The point is not that computers are not capable of telling the difference between a tree and a tank. The point is that a human would not have made this mistake because a human would have understood what it was looking for. Unlike humans, computers do not understand the world.

The cat and the vase

You walk into a room and see a cat sitting on a table and a broken flower vase on the floor. You immediately speculate that the cat may have knocked the vase off the table, where it shattered. Your initial reaction might be wrong. Perhaps a human knocked the vase over and left the room, and the cat just happened to be sitting on the table where the vase used to be. Or perhaps wind coming through an open window blew the vase off the table. Or perhaps an earthquake shook the vase off the table.

You could gather more information to test your theories. Has anyone been in the room and will they confirm they knocked the vase over? Are there any open windows and how strong is the wind outside? Have any earthquakes been reported recently? You might not be able to find a definitive answer, but every theory you consider will make sense.

Could a computer do as well? A computer might observe everything in the room and might even attach correct labels to most things. But, try as it might, for as long as it takes, could it come up with any of the theories that immediately occurred to you? Could it instantaneously discard nonsensical theories that you never considered seriously: the vase jumped off the table onto the floor; the chair reached out and slapped the vase; the rug flew around the room and knocked over the vase?

This is a well-known example of a fundamental difference between humans and machines. Humans can think of theories that make sense, based on logical reasoning and a lifetime of observation. Computers are awful at integrative thinking; for example, using logic, models, and evidence to understand why airplanes fly, why praise is more effective than criticism, why the unemployment rate fluctuates, and why vases fall off tables.

Humans can learn from experience and apply these lessons in other areas. Humans remember seeing animals knock things over and do not remember ever seeing inanimate objects jump about by themselves. Computers are terrible at drawing analogies.

Humans are also very good at anticipating the consequences of everyday events. We have a pretty good idea what will happen if we dive into a cool swimming pool on a hot day, jump off a roof onto a concrete driveway, wave at someone with our hand, kick a soccer ball at a window, plant a tree, ride a bicycle with our eyes closed, smile at a child, yell at our boss. Computers are awful at assessing how one event might cause another.

Computers do not have real-world knowledge—the wisdom and common sense that comes from being alive and storing memories of what we have read, what we have seen, what we have thought about. This is why Big Data and Big Computers can lead to Big Trouble.

CHAPTER 4

Bad data

When I first started teaching economics in 1971, my wife's grandfather ("Popsie") knew that my Ph.D. thesis used Yale's big computer to estimate an extremely complicated economic model. Popsie had bought and sold stocks successfully for decades. He even had his own desk at his broker's office where he could trade gossip and stocks.

Nonetheless, he wanted advice from a 21-year-old kid who had no money and had never bought a single share of stock in his life—me—because I worked with computers. "Ask the computer what it thinks of Schlumberger." "Ask the computer what it thinks of GE."

This naive belief that computers are infallible has been around ever since the first computer was invented more than 100 years ago by Charles Babbage. While a teenager, the great French mathematician Blaise Pascal built a mechanical calculator called the Arithmetique to help his father, a French tax collector. The Arithmetique was a box with visible dials connected to gears hidden inside the box. Each dial had ten digits labeled 0 through 9. When the dial for the 1s column moved from 9 to 0, the dial for the 10s column moved up 1 notch; when the dial for the 10s column moved from 9 to 0, the dial for the 100s column moved up 1 notch; and so on. The Aritmatique could do addition and subtraction, but the dials had to be turned by hand.

Babbage realized that he could convert complex formulas into simple addition-and-subtraction calculations and automate the calculations, so that a mechanical computer would do the calculations perfectly every time, thereby eliminating human error.

Babbage's first design was called the Difference Engine, a steam-powered behemoth made of brass and iron that was 8 feet tall, weighed 15 tons, had 25,000 parts. The Difference Engine could make calculations up to 20 decimals long and it could print formatted tables of results. After a decade tinkering with the design, Babbage began working on plans for a more powerful calculator he called the Analytical Engine. This design had more than 50,000 components, used perforated cards to input instructions and data, and could store up to one thousand 50-digit numbers. The Analytical Engine had a cylindrical "mill" 15 feet tall and 6 feet in diameter that executed instructions sent from a 25-foot long "store." The store was like a modern computer's memory, with the mill the CPU.

Many people were dumbfounded by Babbage's vision. In his autobiography, he recounted that, "On two occasions I have been asked [by members of parliament], 'Pray, Mr. Babbage, if you put into the machine wrong figures, will the right answers come out?'" Babbage confessed that, "I am not able rightly to apprehend the kind of confusion of ideas that could provoke such a question."

Even today, when computers are commonplace, many well-meaning people still cling to the misperception that computers are infallible. The reality, of course, is that if we ask a computer to do something stupid, it will do it perfectly.

Garbage in, garbage out is a snappy reminder of the fact that, no matter how powerful the computer, the value of the output depends on the quality of the input. A variation on this saying is, garbage in, gospel out, referring to the tendency of people to put excessive faith in computer-generated output, without thinking carefully about the input. The truth is that if a computer's calculations are based on bad data, the output is not gospel, but garbage.

There are far too many examples of people worshipping calculations based on misleading data. This chapter gives several examples, but they are just the tip of the statistical iceberg known as Bad Data, which are everywhere. Big Data is not always Better Data.

Humans are capable of recognizing bad data and either taking the flaws into account or discarding the bad data. In carpentry, they say, "Measure twice, cut once." With data, we should, "Think twice, calculate once." Computers cannot do this because, to a computer, numbers are just numbers, and have no real meaning—just as to Nigel Richards, the professional Scrabble player, French words are just letters, and have no real meaning.

As you read through the examples in this chapter, think about whether a computer, no matter how adept at making calculations, has any hope of recognizing bad data.

Self-selection bias

Scientific experiments often involve a treatment group and a control group. For example, 100 tomato plants might be planted in equally fertile soil and given the same water and sunlight. Fifty randomly selected plants (the treatment group) have coffee grounds added to the soil, while the other 50 plants (the control group) do not. We can then see if there are persuasive differences in the plants' health and productivity.

In the behavioral sciences, experiments involving humans are limited. We can't make people lose their jobs, divorce their spouses, or have children and see how they react. Instead, we make do with observational data—observing people who lost their jobs, divorced, or have children. It's very natural to draw conclusions from what we observe, but it's risky.

In 2013, *The Atlantic* published a figure in a bright neon color that made me wish I was wearing sunglasses. It was titled, "The economic value of college, meanwhile, is indisputable." I've spared you the need for sunglasses by using Table 4.1 to show their data.

Table 4.1 *Unemployment and median weekly income. U.S. Adults age 25-plus*

	Unemployment rate, %	Median weekly earnings, $
Doctoral degree	2.5	1,551
Professional degree	2.4	1,665
Master's degree	3.6	1,263
Bachelor's degree	4.9	1,053
Associate's degree	6.8	768
Some college, no degree	8.7	719
High-school diploma	9.4	638
No high-school diploma	14.1	451

A blogger summarized the intended conclusion: "The statistics are very clear: The more education you have, the more money you will make and the less likely you will be unemployed."

I am a college professor and I am not going to dispute the benefits of a college degree, but data like these (which are everywhere) surely exaggerate the financial value of education. People are not randomly assigned to educational categories. People make choices and are screened. Some people do not want to go to college. Some apply to highly selective colleges and are rejected. Some go to college but do not graduate. We can all think of reasons why people have different opportunities and make different choices.

These are examples of the *self-selection bias* that occurs if we compare people who made different choices without considering *why* they made these choices. When people choose to do *what* they are doing, their choices may reflect who they are. This self-selection bias could be avoided by doing a controlled experiment in which people are randomly assigned to groups and told what to do. By taking the choice out of their hands, we eliminate the self-selection bias. Fortunately for all of us, researchers seldom have the power to make us do things we don't want to do simply because they need experimental data.

To make this personal, I was vacationing on Cape Cod in 1983 and a neighbor showed me an article in the *Cape Cod Times*. The Census Bureau regularly estimates the lifetime earnings of people by age, sex, and education. Using these estimates, the *Times* reported that

18-year-old men who receive college education will be "worth" twice as much as peers who don't complete high school. On the average, that first 18-year-old can expect to earn just over $1.3 million while the average high-school dropout can expect to earn about $601,000 between the ages of 18 and 64.

The neighbor said that his son, Mike, was getting poor grades in high school and was thinking of dropping out and going to work for the family's construction firm. But his father read this article and told Mike that dropping out would be a $700,000 mistake. Mike's dad asked me if I agreed.

I tried to answer politely. If Mike was getting poor grades in high school, would he have an even tougher time in college? What kind of job did he want after college? If Mike would be happy working for the family business, he probably wouldn't increase his lifetime income much if he went to college before starting work. If anything, he would be behind financially because of the cost of attending college and the years that he was not working while he was in college.

Mike might get something out of college, but it wasn't likely to be $700,000.

Some college students spend more time partying than studying, which not only lowers their grades but also increases the amount of trouble they get into. One solution is to curb student drinking. In 2001, Harvard University's College Alcohol Study found that students at colleges that ban alcohol were 30 percent less likely to be heavy drinkers and more likely to abstain entirely, suggesting the colleges would benefit greatly from banning alcohol. There may be two kinds of self-selection bias. First, schools with less drinking may be more likely to ban drinking. Second, students who don't drink and don't like drunks may be more likely to choose colleges that ban alcohol.

A variation on this theme is a study of beer drinking in bars near the Virginia Tech campus. This study found that, on average, drinkers who ordered beer by the glass or bottle consumed only half as much as did drinkers who ordered beer by the pitcher. The conclusion, reported nationwide, was that, "If we banned pitchers of beer we would have a significant impact on drinking." The obvious self-selection bias is that people who order pitchers of beer are surely planning to drink a lot and do so. Big drinkers will still be big drinkers even if they are forced to do so by the glass or bottle. Years later, the professor who conducted the study admitted the obvious: many college students "intend to get intoxicated. . . . We have shown in several studies that their intentions influence their behavior. If they intend to get drunk, it's difficult to stop that."

Nonetheless, this study was apparently the inspiration for Stanford administrators to adopt an alcohol policy that prohibits hard alcohol in containers as large or larger than a standard wine bottle: "Limiting the size of hard alcohol containers is a harm reduction strategy designed to reduce the amount of high-volume alcohol content that is available for consumption at a given time. We feel it is a sensible, creative solution that has roots in research-based solutions."

Stanford students immediately recognized the absurdity: people who buy alcohol in large-volume containers presumably plan to consume large amounts of alcohol. If forced to buy smaller-volume containers, they will simply buy more containers. Some students created a mocking web site "Make Stanford Safe Again" with a graphic similar to the one shown below:

For students who do go to class, those who double-major in two subjects generally have higher grade-point averages than do students with only one major—suggesting that students can raise their grades by double majoring. The obvious self-selection bias is that students who choose to double major are systematically different from students who choose not to.

What about life after college? Many students these days go into finance, hoping to make big bucks trading stocks or persuading others to trade stocks. A study from Switzerland's University of St. Gallen concluded that stockbrokers were "significantly more reckless, competitive, and manipulative" than psychopaths. Perhaps stock trading should be banned because of its harmful effects on human behavior? Or perhaps stock trading should be encouraged, since a 1996 study at a Scottish university concluded that, "with the right parenting [psychopaths] can become successful stockbrokers instead of serial killers." Or perhaps there is self-selection bias in that people who are reckless, competitive, and manipulative are attracted to this occupation. Or perhaps the Swiss and Scottish studies were designed to give the results the authors wanted?

For a very different career choice, I attended a Pomona College faculty-trustee event where a representative of the Getty Museum boasted that their internship program, which provides full-time summer jobs at the museum, increased the number of students choosing related careers: "43 percent of our former interns are now working at museums and other nonprofit visual arts institutions." This statistic sounds impressive, but does not prove anything. There is surely self-selection bias, in that those who apply for these internships are likely to be interested in careers in these fields. Did 57 percent of the interns change their minds after doing their internship?

An old chestnut, with no doubt considerable truth, is that people who drive sports cars (especially red sports cars) are more likely to get ticketed than are people who drive minivans. Can you think of a reasonable explanation other than police like to pick on people driving sports cars? The self-selection bias here is that people who choose to drive sports cars may like to drive fast, while people who choose to drive minivans may be concerned about the safety of their children.

Researchers at Northwestern University monitored more than 2,000 adults over the age of 60 and concluded that sitting an extra hour a day caused a 50 percent increase in the chances of having difficulties with daily activities (such as getting dressed by themselves). I am a huge fan of keeping active from birth to death, but perhaps, just perhaps, some people choose to sit because they already have mobility issues?

A study of 20,072 emergency room admissions at a British hospital found that patients who were admitted on public holidays were 48 percent more likely to die within seven days than were patients admitted on other days. One interpretation of these data is that the doctors who work in emergency rooms on public holidays are less qualified and should be avoided. Another possible explanation is that people do not choose to go to emergency rooms on holidays unless it is a life-or-death situation. Or perhaps there are more serious injuries on holidays, due to more travel, excessive drinking, and reckless behavior.

Some places seem to be even more dangerous than emergency rooms. In fact, the most dangerous place to be is apparently in bed, since more people die in bed than anywhere else. Self-selection bias again. People who are near death are often placed in beds, but it isn't the bed that kills them.

Going in the other direction, trying to be healthy, a 1950s study found that married men were in better health than men of the same age who never married or were divorced, suggesting that the healthiest path is for a man to marry and never divorce. However, people who marry and divorce choose to do so. Because of this self-selection bias, the reported data could be true even if marriage is generally bad for a man's health. If women are less likely to marry men who are in poor health, then married men are, on average, in better health than men who never marry. Suppose, too, that women are more likely to divorce men whose health suffers the most during marriage. Then married men are generally in better health than men who have divorced. So, it could paradoxically be the case that marriage is bad for a man's health, yet married men are generally in better health than men who never marry and than men who marry and then divorce. All because of self-selection bias.

A study in one state compared the traffic fatality rates (number of fatalities per miles driven) on highways with 55, 65, and 75 miles per hour speed limits. They found that highways with a 75 mph speed limit had the lowest fatality rate and highways with a 55 mph speed limit had the highest fatality rate. Traffic fatalities could evidently be reduced by raising speed limits, perhaps because people pay more attention when they drive fast. What is the biggest statistical problem with this study? The state surely chose to put 75 mph speed limits on the safest highways (those that are straight, well lit, with light traffic).

In all these cases and many more, a computer program will make erroneous conclusions—predicting that a student's grades will improve after choosing a second major, that a man's health will improve after he marries, that delaying an emergency room visit will increase a seriously injured person's life expectancy, that increasing speed limits will reduce traffic fatalities.

Humans can recognize self-selection bias, computers cannot.

Correlation is not causation

A psychology study found a strong positive correlation between family tension and the number of hours spent watching television, suggesting that watching television increases family tension. Another study found that people who have taken driver-training courses tend to have more traffic accidents than do students who have not taken such courses, suggesting that driver-training courses make people worse drivers. A petition urging schools to offer Latin classes noted that students who take Latin courses get higher scores on tests of verbal ability than do their non-Latin peers.

Statistical software can find such correlations, but it cannot tell us whether the first factor causes the second, the second causes the first, or some third factor causes both. To a computer, numbers are just numbers to be averaged, correlated, and manipulated. Human intelligence gives us the capability to think about the reality behind the numbers and consider plausible interpretations.

Families with a lot of tension may choose to watch television (maybe in different rooms) in order to avoid talking to each other. People who take driver-training classes may do so because they are bad drivers. People who take Latin classes may have more aptitude for and be more interested in the material that appears in tests of verbal ability.

Reverse causation is often related to self-selection bias. When people choose to watch television, take driver-training classes, or take Latin classes, it creates an illusion of causation running from the activity to the person, when the causation may go in the other direction, from the person to the activity.

In 2012, the American Mustache Institute organized a "Million Mustache March" in Washington, D.C., in support of the so-called Stache Act (Stimulus to Allow Critical Hair Expenses), which would give mustached Americans a $250 annual tax deduction for grooming supplies. Their research showed that the mustachioed earned 4.3 percent more per year than the clean-shaven, indicating that a tax policy that encouraged mustache growth would encourage economic growth. They even found a tax-policy professor who endorsed the Stache Act:

Given the clear link between the growing and maintenance of mustaches and income, it appears clear that mustache maintenance costs qualify for and should be considered as a deductible expense related to the production of income under Internal Revenue Code Section 212.

Even if the 4.3 percent number were true (how could it not be made up?), there would clearly be self-selection bias and reverse causation. Is it the mustache-growing that increases people's income or are high-income people more likely to grow mustaches?

For centuries, residents of New Hebrides believed that body lice made a person healthy. This folk wisdom was based on the observation that healthy people often had lice and unhealthy people usually did not. It was not the absence of lice that made people unhealthy, but the fact that unhealthy people often had fevers, which drove the lice away.

Humans can tell the difference between correlation and causation; computers cannot. If I replaced the identifying names of the data in these examples with Xs and Ys, you wouldn't know whether there is a causal relationship. That's what a computer sees—generic data—so how can a computer identify causation? It can't.

That is a fundamental difference between human intelligence and computer intelligence, and one of the reasons why Artificial Intelligence is so often not intelligent at all. To be intelligent, one has to understand what one is talking about.

The power of time

The Pizza Principle says that since the 1960s, the cost of a New York City subway ride has been roughly equal to the cost of a slice of pizza. A computer might conclude that pizza prices depend on subway prices, or vice versa. A human would recognize that both costs have risen with overall prices.

Many things that are otherwise unrelated grow together over time because population, income, and prices all increase over time. This simple fact is the basis for many spurious correlations. The number of married people and the amount of beer consumed have both increased over time. Does marriage cause drinking or does drinking cause marriage? The number of missed days at work and the number of people playing golf have both increased over time. Do people skip work to play golf or do people get injured playing golf? Stork nests and human births increased over time in northwestern Europe, evidently supporting the fairy tale that storks bring babies.

The web site spuriouscorrelations.com has many examples, including the number of lawyers in North Carolina and the number of suicides by hanging, strangulation, and suffocation, and the number of lawyers in Nevada and the number of people who died after tripping over their own two feet. Would reducing the number of lawyers decrease suicides and tripping fatalities?

Sometimes, things decrease over time, like marriage rates and the number of people who die falling out of fishing boats. I once put a graph on an examination showing that U.S. crude oil imports from Norway and the number of automobile drivers killed crossing train tracks both declined between 1999 and 2009, and asked, "How would you explain the 0.95 correlation?" I expected students to say that it was a coincidental correlation between two things that had decreased over time, but were otherwise unrelated. However, one student offered this explanation: "Increased gas prices reduced oil imports from Norway and also reduced driving." Okay, humans may sometimes be fooled by spurious correlations, but computers are *always* fooled.

Not only will a computer program not recognize the correlations discussed here as being bogus, it will actively seek such spurious correlations, and find them. Lacking human intelligence in interpreting what the numbers mean, it cannot assess whether the discovered correlations make sense.

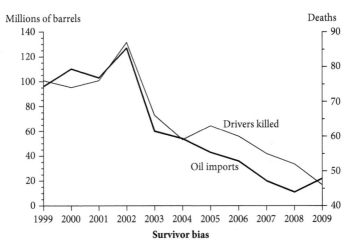

Survivor bias

Survivor bias

Another common problem with observational data is the survivor bias that can occur because we do not see things that no longer exist. A study of the elderly does not include people who did not live long enough to become elderly. A survey of people staying in certain hotels, flying on certain airlines, or visiting certain countries does not include people who did it once and said, "Never again." A compilation of the common characteristics of great companies does not include companies that had these characteristics but did not become great.

In World War II, a military analysis of the location of bullet and shrapnel holes on Allied planes that returned to Britain after bombing runs over Germany found that most of the holes were on the wings and rear of the plane, and very few on the cockpit, engines, or fuel tanks. The military planned to attach protective metal plates to the wings and rear of the aircraft, but a clever statistician named Abraham Wald realized the survivor-bias problem with these data. The planes that returned to Britain were those that survived bullets and shrapnel. There were few holes in the cockpit and fuel tanks because planes that were hit there did not survive. Instead of reinforcing the parts of the planes that had the most holes, they should reinforce the locations with no holes.

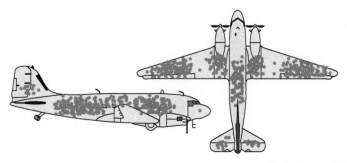

Credit: Cameron Moll

Here is a similar one. Many British soldiers in World War I suffered head injuries from shrapnel. However, when a switch was made from cloth hats to metal helmets, the number of soldiers hospitalized with head injuries increased. Why? Because the metal helmets increased the number of people who were injured instead of being killed by shrapnel.

As with other kinds of bad data, humans can recognize survivor bias, but computer programs cannot, because they do not know what the numbers mean and cannot think logically about biases caused by the source of the data.

Fake data

In the cutthroat world of academic research, brilliant and competitive scientists perpetually seek fame and funding to sustain their careers. This necessary support, in turn, depends on the publication of interesting results in peer-reviewed journals. "Publish or perish" is a brutal fact of university life.

Sometimes, the pressure is so intense that researchers lie and cheat to advance their careers. Needing publishable results to survive, frustrated that their results are not turning out the way they want, and fearful that others will publish similar results first, researchers sometimes take the short cut of making up data. After all, if a theory is true, what harm is there in creating data to prove it?

Diederik Stapel, an extraordinarily productive and successful Dutch social psychologist, was known for being very thorough and conscientious in designing surveys, often with graduate students or colleagues. Oddly enough for a senior researcher, he administered the surveys himself. Another oddity was that Stapel would often learn of a colleague's research interest and claim that he had already collected the data the colleague needed; Stapel supplied the data in return for being listed as a co-author.

Stapel was the author or co-author of hundreds of papers and received a Career Trajectory Award from the Society of Experimental Social Psychology in 2009. He became dean of the Tilburg School of Social and Behavioral Sciences in 2010. Many of Stapel's papers were provocative. Some pushed the boundaries of plausibility. In one paper, he claimed that messy rooms make people racist. In another, he reported that eating meat—indeed, simply thinking about eating meat—makes people selfish. (No, I am not making this up!)

Some of Stapel's colleagues and graduate students were skeptical of how strongly the data supported his half-baked theories and frustrated by Stepel's refusal to let them see his data. In one case, Stapel showed a colleague the means and standard deviations from a study purportedly demonstrating that children who colored a cartoon character shedding a tear were more likely to share candy than were children who colored the same character without tears. When the colleague asked to see the data so that he could compare the responses of boys and girls, Stapel said that the data had not yet been entered into a computer program! Either Stapel had calculated the means and standard deviations by hand, or there were no data. What's your guess?

Finally, two graduate students reported their suspicions to the department chair, and Stapel soon confessed that many of his survey results were either manipulated or completely fabricated. He explained that, "I wanted too much too fast."

Stapel was fired by Tilburg University in 2011. He gave up his PhD and retracted 58 papers (by latest count) in which he had falsified data. He also agreed to do 120 hours of community service and forfeit benefits worth

18 months' salary. In return, Dutch prosecutors agreed not to pursue criminal charges against him for the misuse of public research funds, reasoning that the government grants had been used mainly to pay the salaries of graduate students who did nothing wrong. Meanwhile, the rest of us can feel a little less guilty about eating meat and having messy rooms.

In true capitalist fashion, Stapel did not slink off to meditate on his misdeeds. Instead, he wrote a book about them!

Computers presumably don't make up data, but neither can they recognize fake data when they encounter it. What brought Stapel down was human skepticism of his claims. Computers are not skeptical.

Recognizing bad data

A former student ("Bob") who works with Big Data for the world's largest consulting firm (by dollars) told me about a client company that did a statistical analysis of the efficacy of marketing programs using a variety of channels including direct mail, banner ads, email, and so on. To have enough data to qualify as Big Data, the computer program analyzed more than 20 years of historical data.

The problem, as any human knows, is that marketing technology has changed dramatically in the past 20 years. Consider the Yellow Pages, printed telephone books with a non-commercial directory printed on white pages and paid advertisements organized by categories (such as *Pet Supplies*) printed on yellow pages. These were once so commonplace that, "Let your fingers do the walking!" was part of the national vocabulary. Indeed, the Yellow Pages were so large that tearing one in half was considered a stunning feat of strength (though there is a trick that makes it less challenging).

Now, the Yellow Pages is what Bob calls "a combination coupon book and door stop." Most people who find them in their mailboxes or driveways never open them.

The client's computer software program didn't think about any of this because it doesn't think; it just computes. Fortunately, Bob intervened and convinced the client that a small amount of relevant data is more valuable than a large amount of obsolete data.

It's not just when we look back in time. For all the publicity celebrating Big Data, sometimes Small Data is more useful. Instead of ransacking mountains of essentially random numbers looking for something interesting, it can be more productive to collect good data that are focused on the questions that a study is intended to answer.

Humans have the capacity to be suspicious of data supporting fanciful claims. Computers do not, because they do not have the intelligence to distinguish between the plausible and the fanciful. A computer would not question a statistical relationship between messy rooms and racism, or Nevada lawyers and death by clumsy walking.

One possible solution to the inherent absence of human intelligence in computers is for software programs to report the details of the statistical analysis so that humans can assess the validity of the data and the plausibility of the results. This is a good idea but, in practice, there are two obstacles.

First, because Big Data is so big, many software programs use data-reduction procedures (discussed Chapter 8) that transform the data in ways that make the data unrecognizable. It is no longer possible for humans to assess whether the computer results make sense.

Second, some people don't believe that computer-generated output should be questioned. If a computer says that people who have curved lips are likely to be criminals, then it must be true. If a computer says that people who don't answer their cell phones are bad credit risks, then it must be true.

Complacent humans assume that they are not as smart as computers. Not true. Computers can do some things really, really well, but when it comes to distinguishing between good data and bad data, humans are much smarter than computers—another situation where computer intelligence is not very intelligent.

CHAPTER 5

Patterns in randomness

I do an extra-sensory perception (ESP) experiment on the first day of my statistics classes. I show the students an ordinary coin—sometimes borrowed from a student—and flip the coin ten times. After each flip, I think about the outcome intently while the students try to read my mind. They write their guesses down, and I record the actual flips by circling H or T on a piece of paper that has been designed so that the students cannot tell from the location of my pencil which letter I am circling.

Anyone who guesses all ten flips correctly wins a one-pound box of chocolates from a local gourmet chocolate store. If you want to try this at home, guess my ten coin flips in the stats class I taught in the spring of 2017. My brain waves may still be out there somewhere. Write your guesses down, and we'll see how well you do.

After ten flips, I ask the students to raise their hands and I begin revealing my flips. If a student misses, the hand goes down, Anyone with a hand up at the end wins the chocolates. I had a winner once, which is to be expected since more than a thousand students have played this game.

I don't believe in ESP, so the box of chocolates is not the point of this experiment. I offer the chocolates in order to persuade students to take the test seriously. My real intent is to demonstrate that most people, even bright college students, have a misperception about what coin flips and other random events look like. This misperception fuels our mistaken belief that data patterns uncovered by computers must be meaningful.

Back in the 1930s, the Zenith Radio Corporation broadcast a series of weekly ESP experiments. A "sender" in the radio studio randomly chose a circle or square, analogous to flipping a fair coin, and visualized the shape, hoping that the image would reach listeners hundreds of miles away. After five random draws, listeners were encouraged to mail in their guesses.

These experiments did not support the idea of ESP, but they did provide compelling evidence that people underestimate how frequently patterns appear in random data. Most of us believe that circles and squares should appear about equally often and not in any recognizable pattern. For example, in one experiment, 121 listeners chose this sequence:

1. ▢ ▢ ◯ ▢ ◯

while only 35 chose this sequence:

2. ▢ ◯ ▢ ◯ ▢

Each sequence has three boxes and two circles, but the first sequence seems more random than the perfect alternation of squares and circles in the second sequence. Do you agree?

Only one listener chose this third sequence, because most people do not think that random draws will turn out to be so lopsided.

3. ▢ ▢ ▢ ▢ ▢

The reality is that all three sequences have *exactly* the same probability of occurring. Yet, listeners were reluctant to guess five squares in a row or a perfectly alternating sequence of squares and circles because they didn't think these would happen by chance. Perhaps you feel the same way. BTW, my spring 2017 coin flips were: T, T, T, T, T, H, H, T, H, T. Did you get all ten correct?

After determining whether anyone won the contest, I ask my students to count the longest streak in their coin flips. For this sequence, the longest streak is four heads:

H   T   T   T | H   H   H   H | T   H

For this sequence, the longest streak is three tails:

H   T   T   H   T   H | T   T   T | H

(If these don't seem random, I can assure you they are. I flipped a coin 20 times and this is what I got.)

Of the 263 students who have taken this course over the past ten years, only 13 percent of the students reported streaks of four or longer. Did you?

In fact, a streak of four or more heads or tails in a row is not unlikely at all! In ten fair coin flips, there is a 47 percent chance of a streak of four or longer. We expect 124 of these 263 students to report streaks of four or longer; only 34 did. The students wildly underestimated the chances of four, five, or six heads or tails in a row.

People are evidently very uncomfortable with uninterrupted streaks of heads or tails because such streaks do not seem random. After two or three heads in a row, they feel increasingly compelled to guess tails in order to balance things out.

It is not just coin flips in statistics classes. In sports, games of chance, and life, most people do not appreciate how often streaks appear in random data. So, when streaks do appear, they are too quick to assume that the data must not be random. There must be an underlying cause, so they invent one.

If a basketball player makes five shots in a row, the player is "hot" and he is very likely to make his next shot. A financial adviser who makes five good stock picks must be a financial genius. A mutual fund that has had five good years must be managed by financial geniuses. So, investors move out of funds that have had bad years and into funds that have had good years even though the only consistency in mutual fund performance is that past performance is a poor predictor of future performance.

In his final newspaper column, Melvin Durslag, a member of the National Sportscasters and Sportswriters Hall of Fame, reminisced about advice he had received in his 51 years as a sports columnist, including this suggestion from a famous gambler: "Nick the Greek tipped his secret. He trained himself so that he could stand at the table eight hours at a time without going to the washroom. It was Nick's theory that one in action shouldn't lose the continuity of the dice." Only people who underestimate the likelihood of streaks in random data would struggle to control their bladders while they talk about "the continuity of the dice."

One energetic student was so incredulous of my claim that there is a 47 percent chance that ten coin flips will result in a streak of four or longer, that he wrote a computer program to prove me wrong. His program ran one billion simulations of ten randomized coin flips and recorded the longest streak of heads or tails in each ten-flip simulation. His computer program found that 47 percent of the simulations had streaks of four or longer. He admitted that his program confirmed my claim, but he still was not convinced. He said that something may have been wrong with his computer's random-number generator, but he did not have time to flip a coin a billion times to check. The idea that random numbers don't contain streaks was too ingrained into his thinking.

With more than ten coin flips, the odds are higher that there will be even longer streaks. With 1,000 flips, there is a 62 percent chance of a streak of 10 or longer. With 100,000 flips, there is a 53 percent chance of a streak of 17 or longer and a 32 percent chance of a streak of 18 or longer.

With more data, it is increasingly certain that there will be even longer streaks and other striking patterns. A theoretical paper by Cristian S. Calude and Giuseppe Longo titled, "The Deluge of Spurious Correlations in Big Data," proves that highly regular patterns can be expected in all large data sets. Not only that, but:

the more data, the more arbitrary, meaningless and useless (for future action) correlations will be found in them. Thus, paradoxically, the more information we have, the more difficult is to extract meaning from it. Too much information tends to behave like very little information.

If there is a fixed set of true statistical relationships that are useful for making predictions, the data deluge necessarily increases the ratio of meaningless statistical relationships to true relationships.

Suppose that there is a causal relationship between stock prices, the unemployment rate, and interest rates. When unemployment increases, stock prices tend to fall. When interest rates increase, stock prices tend to fall. If we look at data on stock prices, unemployment, and an interest rate, we are likely to find statistical evidence that confirm these causal relationships.

Now suppose that we also include daily temperatures from several obscure cities that have nothing whatsoever to do with the stock market. As proven by Claude and Longo, the more irrelevant variables we include, the more certain it is that we will find meaningless patterns.

With 2 meaningful variables (unemployment and an interest rate) and 100 meaningless variables (temperatures in 100 small towns), perhaps the 2 meaningful variables and 5 meaningless variables will be closely correlated with stock prices. With 2 meaningful variables and 1,000 meaningless variables, perhaps the 2 meaningful variables and 50 meaningless variables will turn out to be closely correlated with stock prices.

Thus, as Calude and Longo conclude, "The bigger the data, the more likely it is that a discovered pattern is meaningless."

Data mining

Artificial intelligence is an evolving term that encompasses a variety of activities in which computers mimic things that humans do; for example, assembling automobiles, recognizing objects, and translating spoken words into text. AI includes computers that drive cars, play chess, and trade stocks.

The computer programs that control AI activities are called algorithms, step-by-step rules for doing what needs to be done. For example, an algorithm to find the square root of a number might proceed as shown in Table 5.1:

After five loops through the algorithm, the answer comes back: $X = 7.071068$.

Computer programs use various languages to implement algorithms. This square-root algorithm could be written in BASIC, Java, C++, or any

Table 5.1 *Table 1 A square root algorithm*

Rule	Action
1. Input any number Y	$Y = 50$
2. Select a trial solution $X = Y/2$	$X = 50/2 = 25$
3. Calculate X-squared.	$X\text{-squared} = 25*25 = 625$
4. Calculate $Z = Y - X$-squared	$Z = 50 - 625 = -575$
5. Calculate $E = Z/Y$	$E = -575/50 = -11.5$
6. If $|E| < 0.00001$, report X. Otherwise go to step 7	Go to step 7
7. Add $Z/(2X)$ to X	$X = 25 - 575/50 = 13.5$
8. Go back to Step 3	Go to Step 3

other computer language. Artificial intelligence algorithms are of course a lot more ambitious than this simple example.

Data mining is perhaps the most ambitious and dangerous form of artificial intelligence. The traditional statistical analysis of data follows what has come to be known as the scientific method that replaced superstition with scientific knowledge. Based on observation or speculation, the researcher poses a question, such as whether vitamin C reduces the incidence and severity of the common cold. The researcher then gathers data, ideally through a controlled experiment, to test the theory. If there are statistically persuasive differences in the outcomes for those taking vitamin C and those taking a placebo, the study concludes that vitamin C does have a statistically significant effect. The researcher used data to test a theory.

Data mining goes in the other direction, analyzing data without being motivated or encumbered by preconceived theories. Data-mining algorithms are programmed to look for trends, correlations, and other patterns in data. When interesting patterns are found, the researcher may invent a theory to explain the patterns. Alternatively, the researcher might argue that the data speak for themselves, and that is all that needs to be said. We don't need theories—data are sufficient.

In our vitamin C example, a data miner might compile a data base of, say, 1,000 people and record everything we know about them, including gender, age, race, income, hair color, eye color, medical history, exercise habits, and dietary habits. Then data-mining software is used to identify the five personal characteristics that are most closely correlated statistically with each person's number of sick days. These five characteristics might turn out to be high yogurt consumption, low tea consumption, a preference for walking, green eyes, and often using the word *excellent* on Facebook.

The data miner might conclude that yogurt, tea, walking, green eyes, and Facebook *excellent*s are unhealthy, and concoct some fanciful stories to explain these correlations. Or the data miner might believe that the data say all that needs to be said, and no further explanation is needed.

A 2015 article in *The Economist* was titled, "A Long Way From Dismal: Economics Evolves." The article argued that macroeconomists (who study unemployment, inflation, and the like) should follow the lead of those microeconomists who have been hired by tech companies to data-mine data related to products, firms, and markets:

[Macroeconomists] should tone down the theorizing. Macroeconomists are puritans, creating theoretical models before testing them against data. The new breed ignore the white board, chucking numbers together and letting computers spot the patterns.

The Economist is a great magazine, but this was not great journalism.

Back in 2008, Chris Anderson, editor in chief of *Wired*, wrote an article with the provocative title, "The End of Theory: The Data Deluge Makes the Scientific Method Obsolete." Anderson argued that,

With enough data, the numbers speak for themselves. . . . The new availability of huge amounts of data, along with the statistical tools to crunch these numbers, offers a whole new way of understanding the world. Correlation supersedes causation, and science can advance even without coherent models, unified theories, or really any mechanistic explanation at all.

At the time, it seemed to be a deliberately provocative and thinly veiled self-promotion: "The future is big data and big computers; read *Wired*."

To its credit, a few years later, *Wired UK* published a cautionary article, "How to massage statistics," citing my concerns that "computers have made it much easier to fiddle with numbers." and giving a list of ways that data can be massaged, manipulated, and mangled to mislead.

Unfortunately, what was once provocative has become ordinary. Far too many intelligent and well-meaning people believe that number-crunching is enough. We do not need to understand the world. We do not need theories. It is enough to find patterns in data. Computers are really good at that, so we should turn our decision-making over to computers.

Sometimes the term *data mining* is applied more broadly, to encompass such useful and unobjectionable activities as search engines and robotic autoworkers. I use *data mining* specifically to describe the practice of using data to discover statistical relationships that are then used to predict behavior; for example, looking for statistical patterns in order to predict car purchases, loan defaults, illnesses, or changes in stock prices.

Knowledge discovery

I had lunch with a professor who teaches a course called Knowledge Discovery. I asked this professor how, without theory (or, at least, common sense), do we know that patterns in data are useful predictors, and not just coincidental? He argued that,

The proof is in the data. Not only do we not need theories, theorizing restricts what we look at and prevents us from discovering unexpected patterns and relationships. The data tell us all we need to know about whether there is a useful pattern. That is why I call my data-mining course "knowledge discovery."

Data mining has been given a variety of other names, including data exploration, data-driven discovery, knowledge extraction, and information harvesting—all reflecting the core idea that data come before theory, indeed, often without theory.

Much of what is called artificial intelligence is wonderful. Data mining is not. The fundamental problem is simple, but not easily acknowledged:

> We think that patterns are unusual and therefore meaningful.
> In Big Data, patterns are inevitable and therefore meaningless.

Black boxes

I recently reviewed a prospectus for a hedge fund (I'll call it "ThinkNot") that boasted:

Our fully automated portfolio is run using computer algorithms. All trading is conducted through complex computerized systems, eliminating any subjectivity of the manager.

This is called the *black box* approach: inputs are fed into the algorithm, which provides output without human users knowing how the output was determined.

For our square-root algorithm, if the input is 50, the output is 7.071068. My algorithm was not a black box, however, because I explained how the program worked and anyone could check whether I made a mistake, either in my logic or in my step-by-step instructions. In fact, you may have noticed some problems. The square root of 50 can be *plus or minus* 7.071068. My algorithm only showed the positive solution. Also, my program would have a problem with the square root of Y = 0 since Step 5 is supposed to calculate Z/Y, but Z/0 is undefined. Finally, how would my algorithm handle the square root of negative numbers? It couldn't.

When a program is out in the open, humans can see how it works and look for errors, omissions, and other glitches. Not so when the program is hidden inside a black box. When we don't know what is inside a black-box algorithm, we have no way of assessing whether there are logical mistakes, programming errors, or other problems. With black boxes, the inputs are numerous, the process is mysterious, and the output is dubious.

For a black-box stock trading algorithm, the inputs might be data on stock prices, the number of shares traded, interest rates, the unemployment rate, the number of times "stock market" is mentioned in tweets, sales of yellow paint, and dozens of other variables. The output might be a decision to buy or sell 100 shares of Apple stock.

Users who let black-box trading algorithms buy and sell stocks do not know the reason for these decisions, but they are untroubled because they trust the black box, just like Hillary Clinton trusted Ada. They think that their computers are smarter than them. This is meant to be reassuring. Many, like the ThinkNot hedge fund, say that it is a feature not a flaw that the investment decisions are made by a black box, "eliminating any subjectivity of the manager."

A loan-approval black box might reject applicants because they don't keep their phones fully charged. A prison-parole black box might deny parole to applicants who wear wide wristbands. A crime-prevention black box might recommend arresting someone because of the angle of his nose and mouth. You might think I am making this up. I am not.

Computer algorithms do mathematical calculations consistently and perfectly because software engineers know exactly what they want the algorithms to do and write the code that does it. Not so with data mining algorithms, where the intentions are vague and the results are unpredictable. An artificial intelligence expert wrote that, "Any two AI designs might be less similar to one another than you are to a petunia."

Black-box data mining is artificial, but it is not intelligent. That is why I titled this book, *The AI Delusion*.

Some people use the derogatory term *artificial stupidity* (AS) to describe situations in which computers fail us. When Siri does not understand a question. When Google Maps takes us down a dead end road. When an automated traffic light gets stuck on red. I use the term *artificial unintelligence* not to describe the errors that computers sometimes make, but to highlight the fact that computers do not possess the intelligence that humans have. Following rules in order to calculate the square root of 50 is fundamentally different from knowing what the words *Apple stock price* and *Melbourne high temperature* mean, and understanding why there is no logical reason for them to be related.

Big data, big computers, big trouble

Decades ago, when data were scarce and computers nonexistent, researchers worked hard to gather good data and thought carefully before spending hours, even days, on painstaking calculations. Now, we live in the age of Big Data and fast computers, a potent combination that is continually praised—even worshipped—by those who defer to computers as if computers were omniscient. The worship of Big Data has been called *dataism* or *data-ification*, a faith that everything important can be expressed as data and that data analysis is infallible. Bow down to computers.

This is not a harmless obsession. We are too quick to assume that churning through mountains of data can't go wrong. It can. Data are just data. Computers are just computers. Computers cannot distinguish between good data and rubbish. Computers cannot distinguish between sensible conclusions and nonsense. Data without theory is a dangerous philosophy.

Streaks, correlations, trends, and other patterns, by themselves, prove nothing. They can even be found in coin flips. We need to think about reasons. Ask why, not what.

Computers are undeniably wonderful and mysterious. Most of us have no idea how a smartphone allows us to have a video conversation with someone thousands of miles away. How a computer can give you detailed driving directions and the estimated arrival time taking into account current traffic conditions. We just know that computers are amazing. If a computer were to tell us that the outcomes of presidential elections can be predicted by the temperatures in a half dozen cities that we have never heard of, we might be tempted to think that the computer is right. If a computer can tell us the first 2,000 digits of *pi* and show us a street map of every city in the world, who among us mere mortals can question its wisdom?

The harsh truth is that data-mining algorithms are created by mathematicians who often are more interested in mathematical theory than practical reality. Functional magnetic resonance imaging (fMRI) of 15 mathematicians' brains found that looking at mathematical equations activated their medial orbitofrontal cortexes, the same area that is activated when people look at stunning art or listen to wonderful music. Some people appreciate beautiful music, art, dance, literature. Mathematicians appreciate the intrinsic beauty of mathematical equations.

Warren Buffett once warned, "Beware of geeks bearing formulas." I was a math major in college and now I teach finance and statistics. I use math virtually every day of my life and I have written dozens of software programs to analyze data for my research. I love formulas and I love computers, but I also know that the seductive allure of math can lead us to build mathematical models that are intrinsically gratifying, but have no practical value. Too many data-mining algorithms fall into that category.

A conflict of interest

When there is money to be grabbed, people will grab it.

Back in the 1990s, when computers were just starting to take over our lives, the spread of the internet sparked the creation of hundreds of internet-based companies, popularly known as dot-coms. Some dot-coms had good ideas and matured into strong, successful companies. Most did not. In too many cases, the idea was simply to start a company with a *dot-com* in its name, sell it to someone else, and walk away with pockets full of cash. It was so Old Economy to have a great idea, start a company, make it a successful business, and turn it over to your children and grandchildren.

One study found that companies that did nothing more than add *.com*, *.net*, or *internet* to their names more than doubled the price of their stock. Money for nothing!

The same thing is happening now with artificial intelligence. AI has become fashionable and it seems that anything involving computers can be called artificial intelligence. Heck, I could even call my square-root calculator artificial intelligence. Why not?

A former student who invests in AI startups told me that, "Data Scientist and Machine Learning expert are now some of the hottest careers. Some of these people are trained statisticians, some are economists, but some are just programmers who take a 6-week online course, which probably emphasizes some technical tools and tricks but does not provide the theoretical backing to help them understand the theoretical limitations." Who thinks? Call it AI and peddle it. In 2017 the Association of National Advertisers chose "AI" as the Marketing Word of the Year.

Another former student, who is now the chief financial officer of a major company, wrote me that, "You would not believe how frequently the benefits of 'big data' or offers to provide 'analytical expertise' are presented to me—from people who don't know this industry and (likely) are unaware of the limitations detailed in your book."

To persuade people to pay more than they should for something they don't really need, more has to be promised than can realistically be delivered. It happened during the dot-com bubble and it happening now with artificial intelligence. We should be skeptical of claims made by people who are trying to sell us something.

Hard-wired to be deceived

Humans are uncomfortable with the idea of randomness, the thought that things happen for no discernible reason. We try to interpret every pattern as something meaningful, when it may be completely meaningless—just a chance coincidence. As Yogi Berra once said, "It's too much of a coincidence to be a coincidence."

You can blame it on evolution and the environment our distant ancestors grappled with. Living things that have inheritable traits that help them survive and reproduce pass these traits on to future generations, while those with less fortunate traits are weeded out of the gene pool. Continued generation after generation, these valuable inherited traits become dominant.

Recognizing and interpreting patterns used to have survival value. Dark clouds often bring rain. A sound in the brush may be a predator. Hair quality is a sign of fertility. Symmetrical faces are a sign of genetic health. Those distant ancestors who recognized patterns that helped them find food and water, warned them of danger, and attracted them to fertile mates who would breed healthy offspring, passed this aptitude on to future generations. Those who were less adept at recognizing patterns that would help them survive and reproduce had less chance of passing on their genes.

Through countless generations of natural selection, we have become hard-wired to look for patterns and to think of explanations for the patterns we find.

We are too easily seduced by our inherited desire to explain what we see, and this blinds us to the fact that patterns are inevitably created by inexplicable random events—like ten coin flips. We should aspire to recognize our susceptibility to the lure of patterns. Instead of being seduced, we should be skeptical.

Seduced by patterns

The Zenith ESP tests demonstrated that we have pre-conceived (and erroneous) notions of what random data look like. Random data look like sequence #1:

1. ☐ ☐ ○ ☐ ○

Random data don't look like sequence #2:

2. ☐ ○ ☐ ○ ☐

Random data certainly don't look like sequence #3:

3. ☐ ☐ ☐ ☐ ☐

Therefore, we conclude that if a pattern like sequence #2 or #3 occurs, it must not be random. It must be evidence of something real. Perhaps, the squares and circles were not shuffled. Perhaps, it is not an ESP experiment but, rather, a secret code being broadcast to spies.

You may scoff, but Burton Crane, a *New York Times* financial columnist for many years, once reported that:

I have been assured in utter seriousness that the dots between prices on the ticker tape [that used to be used to report stock prices] are a code used by the Big Boys to signal each other for market coups. I have even been shown what purported to be a decoding of the dots.

IBM . . . . . . T. . . . . ADM . . . . . X . . . . ASR . . . . . . GE . . . . . . GM . . . . . .
. . . . 222$3/4$ . . 232$5/8$ . . . . 30$1/2$ . . . 192$1/4$ . . . . . 7$3/8$ . . . . . 331 . . . . 75$1/2$.

Decoding random dots on ticker tape is a primitive form of data mining: find a pattern and make up a theory. Paranoid stock traders scrutinized the dots visually, looking for patterns, found some, and then tried to relate these patterns to changes in stock prices. Motivated to find patterns, traders worked hard to do so—and succeeded. They did not realize that patterns are destined to occur, even in random data.

A variation on this misconception is a book on how to win the dice game Craps. The author recorded the outcomes of 50,000 dice rolls at a Las Vegas casino and studied the sequences in which the numbers appeared. The sequence 4–4–11 can be expected about 20 times in 50,000 roles, but occurred 31 times. The book consequently advised betting on 11 whenever 4 comes up twice in a row. He also found that on 10 of 38 occasions when the sequence 7–12–7 occurred, the next number was either a 2, 3, or 12. A $100 bet on each of these 38 occasions would have won $4,200.

These calculations were all done by hand, before computers, let alone data-mining software. I shudder to think about how many months—maybe years—the author spent searching for patterns. The only consolation is that the more hours this author spent studying the numbers, the fewer hours he spent betting on coincidences.

The hours this unfortunate person wasted discovering coincidental patterns in these 50,000 dice rolls is exactly the kind of thing done today by data mining software, except that computerized data mining can be done in seconds instead of months. These dice rolls are also a simple, easily understood example of how patterns can always be found in data that are as random as dice rolls, and how eager people are to believe that the patterns they discover must be meaningful. The truth is that finding a pattern means nothing at all.

An example of random noise

When big computers ransack Big Data, patterns that are even more complicated and unusual than dice rolls of 4–4–11 are certain to be found even if the data are just random noise. For example, I created 250 "observations" for 100 randomly generated variables. Each variable started at 50 and then my computer's random number generator determined whether the value went up or down for each of the remaining 249 observations. All 100 variables were created by a process that statisticians describe as a *random walk.* Just as a drunkard's next step is unrelated to previous steps, so the next change in each of these variables is unrelated to previous changes.

The evolution of each observation of each variable was completely independent of the other 99 variables. Yet, after the fact, there are bound to be coincidental patterns. Data-mining software is very efficient at finding these patterns, but completely useless in assessing them because, as we have seen over and over again in earlier chapters, computers do not understand the real world. Numbers are just numbers.

I used some data-mining software and found a streak of 13 consecutive up moves in one of these randomly generated variables. If I didn't know better, I might think that I had discovered something important.

Then I used the data-mining software to look for simple pair-wise correlations between any two variables. There are 4,950 possible correlations. My data-mining software found 98 pairs that had correlations above 0.9. If I didn't know better, I might think that I had discovered something important.

Finally, I used the data-mining software to look for combinations of these 100 explanatory variables that might, as a group, be highly correlated with a real variable, the daily values of the S&P 500 index of stock prices in 2015. There are 75,287,520 possible five-variable groups. This sounds like a lot, but not for modern computers. I expected that some five-variable groups of these fake variables would be highly correlated with the real variable, and I was not disappointed. My data-mining software found one combination that has a 0.88 correlation with the S&P 500. If I didn't know better, I might think that I had discovered something important.

In every case, the data-mining software discovered patterns, in one case suggesting that savvy investors could beat the stock market. The software sifted, sorted, and analyzed completely random data which had nothing at all to do with stock prices and which would be completely useless for deciding whether to buy or sell stocks, and it found a correlation strong enough to persuade a black-box stock trading algorithm to buy and sell stocks.

Knowing how the data were created, humans would immediately get the joke. A computer would not. There is no way for data-mining software to know whether it had discovered something useful or useless because, to a computer, numbers are just numbers.

Real data miners who unleash their data-mining algorithms on big data commonly have billions or trillions of observations, and their algorithms not only look for patterns within each data set and the intercorrelations among different data sets, but even more complex relationships. They will inevitably uncover remarkable patterns but, just like this stock market example, the software cannot distinguish between causal and coincidental.

Amateur weather forecasting

Here is another example of the perils of data mining. Data miners routinely sift through data that are tangentially related to what they are trying to predict, even if there is no compelling reason why the data should be of any real value. Suppose, for example, that I want to predict tomorrow's temperature. Real weather forecasters use complicated computer models that divide the atmosphere up into cubes, using satellite data to estimate the temperature, humidity, wind speed, and the like for each cube. Using physics, fluid dynamics, and other scientific principles, the computer models predict how the weather will evolve as the cubes interact with each other.

That sounds like work. I don't have the resources and I don't understand the science. Instead, I used data-mining software to make weather forecasts based on knowledge discovery. Specifically, I tried to predict tomorrow's temperature in City A based on yesterday's temperature in City B. I could use yesterday's temperature in City A, but that wouldn't be knowledge discovery, would it?

I asked a terrific research assistant, Heidi Artigue, to collect the daily high and low temperatures in 2015 and 2016 for 25 widely scattered and relatively obscure U.S. cities. She inadvertently included the Curtin Airport, a small landing strip in Western Australia.

I told you the world is full of coincidences. Several years ago, I visited some friends in Melbourne, Australia, for the Christmas holidays. I learned about plum pudding, Aussie Jingle Bells, and the Boxing Day Test at the Melbourne Cricket Ground. I even played backyard cricket with tennis balls. The most memorable moment, however, was during the opening of Christmas presents. Two brothers gave their elderly mum round-trip tickets to Perth on the western side of Australia. She opened the envelope, squinted at the tickets, frowned, and complained loudly, "Why the hell would I want to go to Perth?" She had lived in Melbourne, which is in Eastern Australia, her entire life and she had no interest in flying across the country for a vacation in Western Australia.

In honor of that memory, I chose Curtin as my forecast city and used data-mining software to see how well I could forecast the daily low temperature in Curtin by using the high and low temperatures for 24 equally obscure U.S. cities. My data mining turned up Omak, Washington, a U.S. city with cold winters, hot summers, and fewer than 5,000 residents. Its daily *high* temperature happened to have a -0.77 correlation with the next day's *low* temperature at the Curtin Airport in Australia.

The correlation between the temperature in Omak and the temperature the next day in Curtin is negative because Omak is in the Northern Hemisphere and Curtin is in the Southern Hemisphere. The -0.77 correlation is very impressive considering that the two cities are on the opposite sides of the earth. Figure 1 shows a scatter plot.

A mindless data-mining program (and all data-mining programs are mindless) might conclude that this is a knowledge discovery of a useful tool for forecasting the weather in Curtin. A mindful human would think that it is preposterous to think that the best way to forecast the

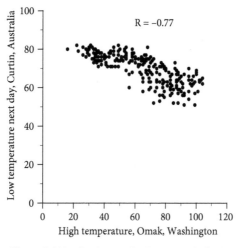

Figure 1 Using Omak to predict the weather in Curtin

low temperature tomorrow in a town in Australia is to look at the high temperature today in a town in Washington.

I unleashed my data-mining software on another collection of data and easily found an even closer correlation. Figure 2 shows a 0.81 correlation between the daily low temperature in Curtin, Australia, and Random Variable 58. Yep, the variable on the horizontal axis is one of the 100 variables that I created using my computer's random number generator to forecast stock prices.

These phony variables were generated completely independently of the weather in Curtin and yet I found one (Random Variable 58) that happened to be closely correlated with Curtin's weather. That's the thing about coin flips and other random noise. They often result in patterns and correlations that look real but are meaningless.

I only looked at 100 random variables. With modern computers, it would be easy for me to look at thousands or millions of random variables until I stumble upon one with an astonishingly close correlation with the temperature in Curtin, or any other city for that matter.

And what, exactly, would I prove? Nothing at all—and that is the first thing to remember about data mining. If we scrutinize lots of data, we

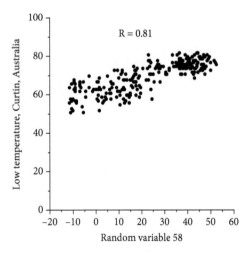

Figure 2 Tapping the power of random selection

will find statistical patterns no matter whether something real is going on or not. The second thing to remember is that, even though it is called artificial intelligence, data-mining software is not intelligent enough to tell the difference between patterns that reflect real relationships and patterns that are coincidental. Only humans can do that.

The Smith test

Suppose that a data-mining algorithm discovered that U.S. stock prices are correlated with the daily low temperature in Curtin, Australia. How would the computer program know if this statistical relationship is real or coincidental? In contrast, humans know what stock prices and temperatures are. They also know stocks are not made more or less valuable by temperatures in Curtin, Australia.

A computer could retrieve a definition of *stock*, though it might be a different kind of stock; perhaps merchandise, animals, or bouillon. Even if it finds the correct definition, a computer does not know what the words in the definition mean, though it could retrieve definitions of these words, too, and then definitions of the words in those definitions. Beyond retrieving definitions, a computer does not know, in any real sense, what a stock is, what a stock exchange is, what a stock price is, or why stock prices go up and down. Nor does it know what low temperatures in Curtin are or why they may or may not be related to U.S. stock prices.

A computer program could search through a data base of published research and look for articles that mention stock prices and Australian temperatures. However, it is exceedingly difficult (impossible?) for computers to interpret the relevance of research that happens to involve these words. It is also difficult for a computer to evaluate whether the research is valid. John Ioannidis has argued persuasively that most published medical research is wrong, including research published in the most prestigious medical journals (often because the reported results were data mined). I believe that the same is true of most stock market research. We will look at the reasoning behind these arguments in later chapters; for now, the point is simply that a computer search for the words *stock prices* and *Australian temperatures* is unlikely to turn up anything that the computer would interpret as supporting or contradicting the statistical pattern it discovered, and, if it did find anything, the computer would be hard-pressed to assess the reliability of what it found.

In addition, the whole claim of "knowledge discovery" is that computers will discover new, previously unknown patterns and relationships. By definition, a knowledge discovery is not something that has already been published. How can a computer without wisdom or common sense tell whether its knowledge discovery makes sense? It cannot. Computers do not have wisdom or common sense.

We are back in the Chinese Room. If a computer does not truly understand the reality represented by the words *stock prices* and *temperatures*, it cannot tell whether the statistical patterns it discovers are meaningful or coincidental. Call it theoretical knowledge, human intuition, experience, wisdom, or common sense, but there is a fundamental difference between a computer that discovers statistical relationships by looking at data and a human who anticipates relationships without looking at the data.

I propose what I modestly call the Smith test:

Collect 100 sets of data; for example, data on U.S. stock prices, unemployment, interest rates, rice prices, sales of blue paint in New Zealand, and temperatures in Curtin, Australia. Allow the computer to analyze the data in any way it wants, and then report the statistical relationships that it thinks might be useful for making predictions. The computer passes the Smith test if a human panel concurs that the relationships selected by the computer make sense.

There can be true knowledge discovery in that the computer may suggest relationships that the humans overlooked, but consider plausible. The computer fails, however, if it selects relationships that humans consider nonsensical, such as U.S. stock prices and temperatures in Curtin, Australia.

CHAPTER 6

If you torture the data long enough

I recently received an e-mail that offered me a way to automate my research:

Dear Professor Smith,

We would like to introduce you to [our] brand new research tool . . . , ready to automate your empirical research basing on official statistical time series databases. [Our software] has been designed to explore and discover new exciting economic correlations directly from your desktop.

No extra software required, no need to crawl thousands of databases manually. You'll be up and running in no time your first big data project.

The e-mail went on to boast that their software will calculate "correlation coefficients with millions of statistical time series," "identify unexpected interdependences," and "find new insights."

The creative grammar was one thing. More disheartening was their assumption that I wanted to sift through literally trillions of correlations looking for unexpected patterns. An unexpected pattern has no logical basis—and I am skeptical of patterns that defy logic.

Statistical tests assume that researchers have well-defined theories in mind and gather appropriate data to test their theories. This company assumed that I was eager and willing to pay a substantial amount of money to work the other way around. Look at every possible correlation—not caring whether they made sense or not—and report the correlations that turn out to be the most statistically persuasive.

It is a sign of the times, but not an inspiring sign.

Mendel

Many important scientific theories started out as efforts to explain observed patterns. For example, during the 1800s, most biologists believed that parental characteristics were averaged together to determine the characteristics of their offspring. For example, a child's height is an average of the father's and mother's heights, modified by environmental influences.

However, Gregor Mendel discovered something quite different in his experiments with pea plants. Mendel was born in Austria in 1822 and grew up on his family's farm. His parents expected him to take over the farm, but Mendel was an excellent student and became an Augustinian monk at a monastery known for its scientific library and research.

Perhaps because of his farming roots, Mendel conducted meticulous studies of tens of thousands of pea plants grown in the monastery's gardens over an eight-year period. He looked at several different traits and concluded that the blending theory could not explain the results of his experiments. When he cross-pollinated yellow-seeded plants with green-seeded plants, the offspring's seeds were green or yellow, not a yellowish-green mixture of the parental seeds, nor were they equally likely to be green or yellow. Similarly, when he cross-pollinated smooth-seeded plants with wrinkly-seeded plants, the offspring's seeds were smooth or wrinkled, not something in between, and again not equally likely to be smooth or wrinkled.

To explain the results of his experiments, he proposed what are now known as Mendel's laws of inheritance. He concluded that genes can exist in more than one form (now called *alleles*); for example, an allele for yellow seed color or an allele for green seed color. Offspring have a pair of alleles for each trait, one allele independently inherited from each parent, and the inherited alleles may be the same (homozygous) or different (heterozygous). In his experiments, pea plants that were heterozygous for seed color had yellow seeds; so, yellow is the dominant allele and green is recessive. Seeds are green only if both alleles are recessive.

Mendel carefully constructed a theory that not only fit his data, but made sense. Had they existed at the time, computers could not have done as well. We don't observe alleles, so computers would not have had the inspiration to assume that they exist, that each parent has two, that the offspring inherit one from each parent, and that an allele may be dominant or recessive.

We now know that some traits do not exhibit complete dominance, with the dominant allele determining the trait. When there is incomplete dominance, traits are a blend; for example, the heterozygous offspring of red-flowered snapdragons and white-flowered snapdragons have pink flowers. Tulip flowers exhibit co-dominance in that the heterozygous offspring of red-flowered tulips and white-flowered tulips are both red and white.

Mendel created a theory to fit his data and thereby laid the foundation for modern genetics. However, the mantra "data first, theory later" has also been the source of thousands of quack theories. Have you heard the one about baseball players whose first names began with the letter D dying years younger than do players whose first names began with the letters E through Z? Or that Asians are prone to heart attacks on the fourth day of every month?

The Texas sharpshooter fallacy

Two endemic problems with "data first, theory later" are nicely summarized by the Texas sharpshooter fallacy.

1 In one version, a self-proclaimed marksman covers a wall with targets and then fires his gun. Inevitably, he hits a target, which he displays proudly without mentioning all the missed targets. Because he is certain to hit a target, the fact that he did so proves nothing at all. In research, this corresponds to testing hundreds (or thousands or millions) of theories and reporting the most statistically persuasive results, without mentioning all the failed tests. In extrasensory perception (ESP) research, for example, someone might administer dozens of tests to thousands of subjects and only report those tests (or parts of tests) that support the ESP claim. This proves nothing because the researcher is certain to find supportive evidence if enough tests are conducted.

2 In the second version of the sharpshooter fallacy, the hapless cowboy shoots a bullet at a blank wall. He then draws a bullseye around the bullet hole, which again proves nothing because there will always be a hole to draw a circle around. The research equivalent is to ransack data for a pattern and, after one is found, think up a theory. In ESP research, someone might report that although a subject's responses did not match the cues being recorded at that time, they did match the cues

recorded earlier ("backward displacement"), recorded later ("forward displacement"), or not recorded ("negative ESP"). Someone who looks for a pattern will inevitably find one. So, finding something only proves someone looked.

The Texas sharpshooter fallacies are known by a variety of names, including data mining, data dredging, fishing expeditions, cherry picking, data snooping, and p-hacking. The p-hacking nickname comes from the fact that test results are considered statistically significant if there is a low probability (p value) that they would occur by chance. A Texas sharpshooter has a high probability of reporting a low p-value; so he is a p-hacker.

The journals that publish research exacerbate this situation because they prefer (or demand) statistically significant results, which rewards researchers for using the sharpshooter fallacies to obtain statistically significant results.

Fallacy #2 is also called the Feynman Trap, a reference to Nobel Laureate Richard Feynman. Feynman asked his Caltech students to calculate the probability that, if he walked outside the classroom, the first car in the parking lot would have a specific license plate, say 8NSR26. The students calculated a probability by assuming each number and letter were equally likely and independently determined. The answer is less than 1 in 17 million. When the students finished their calculations, Feynman revealed that the correct probability was 1 because he had seen this license plate on his way to class. Something extremely unlikely is not unlikely at all if it has already happened.

A concise summary of the sharpshooter fallacies is the cynical comment of Nobel Laureate Ronald Coase: "If you torture the data long enough, it will confess."

I will use the generic term *data mining* to encompass the two Texas sharpshooter fallacies and related mischief.

Data mining

Decades ago, when I was just starting my career, being called a "data miner" was an insult comparable to being accused of plagiarism. If some-one presented results that seemed theoretically implausible or too good to be true (for example, a near perfect correlation), a rude retort was, "Data mining!"

I remember one seminar where a visiting professor claimed that there was a close empirical relationship between the U S. economy and a bizarre measure of the money supply that he had created so that it would be closely correlated with the U. S. economy. The close correlation proved nothing at all because he had created the close correlation.

The audience was unusually restrained, perhaps because of the visitor's status, but as we left the seminar room, many murmured, "data mining." It was not meant to be a compliment. Years later, I re-examined his model and was not surprised to find that, while there was a close relationship during the historical period he used to concoct his goofy measure of the money supply, there was a lousy fit afterward—because there was no theoretical basis for his model. He had been a data miner and his model was worthless.

Today, people advertise themselves as data miners and charge unconscionable fees for mechanical calculations done by mindless computers. The e-mail that invited me to sift through millions of data sets looking for unexpected statistical relationships assumed that I would pay to join their ranks.

Data miners used to be the exception. James Tobin, a Nobel laureate in economics, wryly observed that the bad old days when researchers had to do calculations by hand were actually a blessing. In today's language, it was a feature, not a flaw. The calculations were so hard that people thought hard before calculating. Today, with terabytes of data and lightning-fast computers, it is too easy to calculate first, think later. This is a flaw, not a feature. It is better to think hard before calculating.

Who thinks? Just let the software ransack the data looking for unexpected relationships.

QuickStop

An internet marketer (QuickStop) owned more than a million domain names that it used to lure inadvertent web traffic. For a hypothetical example, someone who was looking for the Hardle Soup Company (which is at www.hardlesoup.com) and mistakenly typed www.hardle.com or www.hardlysoup.com was sent to a QuickStop *landing page* (or *lander*), where the company peddled products.

QuickStop's senior manager thought the company might boost its revenue if it changed its traditional blue landing-page color to green, red, or teal. The company's data analysts set up four versions of the landing page and, after several weeks of tests, concluded that none of the three alternative colors led to a significant increase in revenue. The senior manager was disappointed (it was his idea, after all) and suggested that they might find some differences if they separated the data by country. Maybe people prefer the colors in their national flags or colors that match their nation's personality?

Sure enough, when the data were broken down by country, they found one nation where color mattered: England loves teal. Great Britain is an island and maybe the teal color reminded the British of the waters surrounding Great Britain? The manager said that it didn't really matter why; the important point is that the English spend more when the landing page is teal.

QuickStop was going to switch its British landing page to teal when a QuickStop data scientist blew an imaginary whistle and called a "data foul." This was pure data mining. QuickStop hadn't initiated the experiment with the thought that the British are the only country where the color of the landing page matters, let alone that teal would be the preferred color. By looking at three alternative colors for 100 or so countries, they virtually guaranteed that they would find a revenue pop for some color for some country. The more colors and the more countries, the more likely it is that—by chance alone—they would find something significant.

The data scientist insisted that they continue the test and gather additional British data. He wasn't surprised when the teal effect disappeared. In fact, revenue was lower with the teal landing page than with the company's current blue color.

That was Texas Sharpshooter Fallacy #1: test lots of theories and focus on the test that finds the most significant result. Here's an example of Fallacy #2. Another internet marketing company (TryAnything) gave free samples to people who filled out a questionnaire on its landing page. TryAnything then unleashed a data-mining program to find correlations between the questionnaire answers and the people who ordered the product after receiving the sample. The software found a surprising relationship between purchases and the length of a person's last name—people with more than nine characters in their last name were more likely to order the product. TryAnything's executives speculated that this might have something to do with ethnicity. They hadn't asked any questions about ethnicity for fear of offending some people, but perhaps name length was a proxy for ethnicity, and their product appealed to some groups more than others. Or maybe not. TryAnything's management figured that they didn't need an explanation. They had expensive software and a powerful computer and that was enough. The data are what they are.

If QuickStop's skeptical data scientist had been working for TryAnything, he would have called a data foul. The data-mining program had no particular theories in mind when it sliced and diced the questionnaire answers. The questionnaires had dozen of entries and some answers—by chance alone—were bound to be correlated with product sales. But Quick-Stop's data scientist was not working for TryAnything, so the company was not hit with a data foul. TryAnything identified all the respondents who had long last names and had its sales people make personal telephone calls to each. No robocalls. Real calls by real people who were being paid real money to make the calls. It was a colossal flop. The personal calls were a money pit and were abandoned after a few months.

Torturing data

Do serious researchers really torture data? Far too often. It's how well-respected people came up with the now-discredited ideas that coffee causes pancreatic cancer, that people can be healed by positive energy from healers living thousands of miles away, and that hurricanes are deadlier if they have female names.

Nick Brown, a master's student in psychology, reported that an advanced graduate student gave him what was intended to be helpful advice: "You want to know how it works? We have a bunch of half-baked ideas. We run a bunch of experiments. Whatever data we get, we pretend that's what we were looking for."

This cavalier attitude and the pressure to obtain publishable results by any means possible is why so much published research is tenuous, flimsy, or outright junk. The Reproducibility Project attempted to replicate 100 studies published in three prestigious cognitive and social psychology journals and could only replicate 36. We will look at similar dismal statistics for medical journals in a later chapter. These were the best studies in the best journals. Imagine the statistics for lesser journals.

The biggest distinction between academic and commercial research is that academics get fame and funding by publishing their research, while commercial researchers usually keep their studies confidential because they are potentially profitable proprietary information. Google doesn't share its search algorithms. Goldman Sachs doesn't share its stock-trading algorithms.

However, the people doing academic and commercial research are similar. In fact, many people move back and forth between the two: I do academic work that gets published; I do consulting work that doesn't. I am hardly alone.

Academic and commercial researchers receive similar training; have the same toolbox of techniques; and are subject to similar pressures. Academics are rewarded for coming up with interesting empirical findings; so are commercial researchers. Academics who don't have anything new and interesting to report are overlooked and underfunded, but at least they have tenure. Commercial researchers who find nothing are fired.

While most examples of commercial researchers torturing data are hidden by proprietary protocol, there are many examples of tortured research in academic journals. Let's look at a few for insights into how data can be abused.

Retroactive recall

Some researchers are not only not unapologetic about torturing data, they even encourage it. Daryl Bem, a prominent social psychologist wrote that,

The conventional view of the research process is that we first derive a set of hypotheses from a theory, design and conduct a study to test these hypotheses, analyze the data to see if they were confirmed or disconfirmed, and then chronicle this sequence of events in the journal article. . . . But this is not how our enterprise actually proceeds. Psychology is more exciting than that.

He goes on:

Examine [the data] from every angle. Analyze the sexes separately. Make up new composite indexes. If a datum suggests a new hypothesis, try to find further evidence for it elsewhere in the data. If you see dim traces of interesting patterns, try to reorganize the data to bring them into bolder relief. If there are participants you don't like, or trials, observers, or interviewers who gave you anomalous results, place them aside temporarily and see if any coherent patterns emerge. Go on a fishing expedition for something— anything—interesting.

Using the fishing-expedition approach, Bem was able to discover evidence of some truly incredible things. In a 2011 paper titled, "Feeling the Future," Bem reported that when erotic pictures were shown in random locations on a computer screen, his subjects were able to guess beforehand, with 53 percent accuracy, whether the picture would be on the left or right side of the screen. Part of Bem's fishing expedition was that he did his experiment with five different kinds of pictures, and chose to emphasize the only type that was statistically significant. It is striking that, even using Sharpshooter Fallacy #1, the best he could do was 53 percent, barely better than a coin flip.

An even more astonishing claim by Bem was that people have "retroactive recall." Volunteer Cornell undergraduates were shown 48 common nouns on a computer screen, one at a time for three seconds each. The students were instructed to visualize the object that the word referred to (for example, visualizing a tree if the word was *tree*). After seeing all 48 words, the students were instructed to type all the words they could remember. After they finished, the subjects were shown 24 randomly selected words from the original 48 and asked to do various tasks that might help them remember the words; for example, clicking on words related to trees and retyping these words. Bem reported that the students were more likely to remember words during the recall test if they studied these words *after* they took the recall test. For example, if they manipulated the word *tree* after taking the recall test, this increased the probability that they remembered the word *tree* during the recall test.

Imagine the implications. Cornell students who are pressed for time during the semester can study for their examinations after the semester is over and they have already taken their tests. Internet advertisers can increase sales by running ads after people have decided whether or not to buy something.

As might be expected, other researchers were unable to replicate Bem's results. The journal that published his "Feeling the Future" paper, published a paper a year later titled "Correcting the Past," coauthored by four professors at four different universities. They reported that seven experiments attempting to replicate Bem's claim that people can feel the future "found no evidence supporting its existence." This damning refutation of Bem's far-fetched claims added considerable fuel to the broader question of how such papers get published in reputable journals. It was also revealed that other attempts to publish studies that refuted Bem's claims were rejected by journals that have a policy of not publishing replication studies or, in one case, because a referee who recommended rejection was Bem.

Money priming

I was invited to Sci Foo 2015, an annual gathering of around 250 scientists, writers, and policy makers at the Googleplex, Google's corporate headquarters in Mountain View, California. On the opening night, I talked to a social psychologist who told me about some interesting experiments. Polish children ages 3 to 6 ("a population that does not understand money") were separated into two groups. One group sorted paper money by color; the other sorted buttons.

The children then went into a different room and were asked to solve a maze. The money sorters worked longer and were more successful ("better work outcomes"). In another experiment, the children were asked to go across the room and bring back red crayons. I guessed, based on the results of the first experiment, that the money sorters would bring back more crayons, since they seemed to like working. But, no, the money sorters brought back fewer crayons, which the researchers interpreted as evidence that the money-sorting children were less helpful.

In their published paper, the researchers concluded that handling money "increased laborious effort and reduced helpfulness and generosity," and these effects "were not contingent on the money's value, the children's knowledge about money, or their age."

My initial reaction was skepticism about the whole enterprise. How could 3-year-olds who don't know what money is be affected by touching money? My doubts were heightened by the interpretation of the red-crayon results. If the money-sorters had brought back more crayons, this might have been interpreted as additional evidence that they were better workers. When they brought back fewer crayons, I would have interpreted it as evidence contradicting the claim that they were better workers; instead, it was conveniently interpreted as evidence that they were less helpful. It sure seemed like Texas Sharpshooter Fallacy #2 to me.

I asked this social psychologist what she thought of Bem's research. She praised it. I pressed on by asking her what she thought of Diederik Stapel, the social psychologist who had admitted that he made up data, and she praised him too. I changed the subject.

The next day at the Googleplex, I heard chemists, biologists, astrophysicists, and other scientists express their concerns about the "replication crisis," fanciful claims that are published in reputable journals, but cannot be replicated, and how this is undermining the credibility of scientific research. One prominent social psychologist said that his field is the poster child for irreproducible research, and that his default assumption is that anything published in his field is wrong, evidently because too many social psychologists do not understand the fundamental truth that discovering a pattern in ransacked data proves nothing more than that they ransacked the data looking for patterns. I also overheard a group talking about the implausibility of the claim that children are affected by seeing or handling money.

When I got back home, I poked around the internet looking for other money studies and I discovered a rich literature on "priming," which analyzes how our short-term memory links our responses to a series of stimuli. For example, if subjects are asked to say a word that begins with *rep*, they are likely to say *repair* if they have recently seen that word. Similarly, if asked whether *rebuild* is a word, they answer faster if the preceding word is *repair* than if it is *swim*. These conclusions make sense.

Less sensible are money priming claims, like young children working harder and becoming more selfish after touching money. I found money-priming papers claiming that adult behavior is influenced by seeing sentences related to money (*we can afford it*) or seeing computer backgrounds that included faint images of money. One paper (co-authored by the social psychologist I chatted with at the Googleplex) has the provocative title, "Mere exposure to money increases endorsement of free-market systems and social inequality." The authors found that,

subtle reminders of the concept of money, relative to nonmoney concepts, led participants to endorse more strongly the existing social system in the United States in general (Experiment 1) and free-market capitalism in particular (Experiment 4), to assert more strongly that victims deserve their fate (Experiment 2), and to believe more strongly that socially advantaged groups should dominate socially disadvantaged groups (Experiment 3).

Strong stuff! So strong that I'm pretty sure it isn't true.

Then I found an ambitious attempt to replicate this paper. It failed to find any of the effects that were reported. The authors suggested that Texas Sharpshooter Fallacy #1 played a role in that the original paper neglected to mention several tests that did not support their conclusions. The money-priming authors fired back, which prompted a third group of scientists to enter the fray and reach an even more damning conclusion: the claimed money-priming effects "are distorted by selection bias, reporting biases, or p-hacking."

Seek and you will find

I was living on Cape Cod in 1991 when Hurricane Bob hit and it seemed like a cruel joke that one of the most devastating hurricanes in New England history would have such an innocent name. Little did I anticipate that, years later, researchers at the University of Illinois would claim that hurricane names do matter.

When a reporter first asked me what I thought about a study entitled, "Female Hurricanes are Deadlier than Male Hurricanes," I was skeptical. Hurricane names have alternated between male and female since 1979 and it seemed implausible that a correlation between a hurricane's name and its death toll would be anything other than coincidence.

It turns out that, despite the study's title, the authors' argument is not that female-named hurricanes are stronger, but that people don't take hurricanes with female names seriously and, as a deadly consequence, don't prepare properly and are more likely be killed. Their study was published in the *Proceedings of the National Academy of Sciences (PNAS)*, so it must be correct, right? Not necessarily. Lots of junk science has been published in prestigious journals.

When I looked at the details of the study. I found several compelling reasons for skepticism.

In 1950, 1951, and 1952, hurricanes were named using the military phonetic alphabet (Able, Baker, Charlie, . . .). A switch was made to all female names in 1953. Many feminists decried this sexism, with Roxcy Bolton noting that, "Women are not disasters, destroying life and communities and leaving a lasting and devastating effect." The switch to the current system of alternating male and female names was made in 1979.

The *PNAS* study included pre-1979 data, when all hurricanes were given female names. The inclusion of these pre-1979 data is problematic because the average number of deaths per hurricane was 29 during the all-female era and 16 afterward. The overall female average was consequently boosted by the fact that hurricanes were deadlier before 1979, when all hurricane names were female. Perhaps there were more fatalities in earlier years because hurricanes tended to be stronger (the average hurricane category was 2.26 during the all-female era and 1.96 afterward), the infrastructure was weaker, or there was less advance warning.

There is no concrete way to compare storm warnings before and after 1979, but there is anecdotal evidence that there is more advance notice now. On September 20, 1938, the *Springfield Union* newspaper in Springfield, Massachusetts, printed this weather forecast for western Massachusetts: "Rain today and possibly tomorrow." The Great Hurricane of 1938 hit the next day, killing 99 people in Massachusetts. In Springfield, the Connecticut River rose six-to-ten feet above flood level. Overall, nearly 700 people were killed and property damage was estimated at nearly $5 billion in 2015 dollars. Seventy-five years later, the chief National Weather Service Meteorologist in Taunton, Massachusetts, observed that, "It is inconceivable for a hurricane to arrive unannounced like it did in 1938."

The National Oceanic and Atmospheric Administration (2012) boasted that,

NOAA's investment in ocean and atmospheric research, coupled with technological advancements, has led to a remarkable transformation in hurricane monitoring and forecasting. Emerging from these combined factors has come intricate computer modeling, a vast network of ground- and ocean-based sensors, satellites, and Hurricane Hunter aircraft.... Advances of the last half-century have brought tremendous improvements in hurricane forecasting and, despite a growing coastal population, have yielded a dramatic decline in hurricane-related fatalities.

Even allowing for some self-promotion by the NOAA, it is clearly potentially misleading to treat the storm danger before 1979 the same as in more recent years. It is more scientifically valid to analyze storms since 1979, when male and female names have alternated.

The *PNAS* study also omitted several deadly hurricanes. One account described Hurricane Bill in 2009:

Large swells, high surf, and rip currents generated by Bill caused two deaths in the United States. Although warnings about the dangerous waves had been posted along the coast, over 10 000 people gathered along the shore in Acadia National Park, Maine, on 24 August to witness the event. One wave swept more than 20 people into the ocean; 11 people were sent to the hospital, and a 7-yr-old girl died. Elsewhere, a 54-yr-old swimmer died after he was washed ashore by large waves and found unconscious in New Smyrna Beach, Florida.

Hurricane Bill was not included in the *PNAS* study's data because the hurricane didn't quite make it to shore. But, surely, this episode is evidence that people didn't take Hurricane Bill seriously.

The authors wrote that in the data they did analyze, female names "cause" more deaths during major storms but that there is "no effect of masculinity-femininity of name for less severe storms." That seems backwards. The difference between male and female-named storms should be smaller for major storms. Consider Hurricane Sandy, the deadliest post-1978 hurricane in their data set.

To begin with, Sandy is generally considered a unisex name, but the *PNAS* authors considered it to be strongly feminine. They asked nine people to gauge the masculinity or femininity of hurricane names on a scale of 1 to 11 and reported that Sandy got an average score of 9.0 (strongly feminine)—more feminine than Edith (8.5), Carol (8.1), or Beulah (7.3). Something is surely amiss. I surveyed 44 people and got an average score of 7.25, which is more plausible.

Hurricane Sandy made landfall in Jamaica on October 24, 2012, killing 2 people and knocking out 70 percent of the island's power, and made landfall in Cuba on October 26, with winds of 155 miles per hour, killing 11 people and destroying more than 15,000 homes. An additional 2 people were killed in the Dominican Republic, 54 in Haiti, and 1 in Puerto Rico. Nine U.S. governors declared a state of emergency before Sandy made landfall in New Jersey on October 29. New York City Mayor Michael Bloomberg suspended all city mass transit services, including busses, subways, and trains; closed public schools; and ordered mandatory evacuations of many parts of the city.

Nonetheless, there were 48 New York City fatalities and another 109 fatalities in other parts of the United States. Is it really credible that, despite the dozens of people killed before Sandy hit the United States and the extraordinary actions taken by elected US officials, people did not take Hurricane Sandy seriously because they considered Sandy to be a female name?

If the implicit-sexism theory is true, it ought to be most apparent for storms of questionable danger. It is implausible that the response to a potential storm of the century—with the catastrophic warnings broadcast by news media—depends on whether the name is perceived to be feminine or masculine. It is more plausible that relatively minor storms (like Hurricane Bill) might be dismissed as more nuisance than danger.

Interestingly, Sandy was downgraded from a hurricane to a posttropical cyclone when it made landfall in New Jersey. Should the *PNAS* authors have included deadly hurricanes (like Bill) that did not make landfall? Should they have excluded storms (like Sandy) that were no longer hurricanes when they made landfall? Should they have included storms that were deadly, but didn't reach hurricane status? Who knows? And that's the point. The authors also reported that they estimated a large number of models with various combinations of variables (Texas Sharpshooter Fallacy #1).

When a strong, surprising conclusion is drawn from tortured data, it can be instructive to see whether the conclusion is robust with respect to the myriad decisions used to torture the data. I tested the robustness of the reported results using a more inclusive set of post-1978 data: tropical storms as well as hurricanes, Pacific storms, storms that did not make landfall or made landfall in other countries, and non-U.S. fatalities.

Tables 6.1 and 2 show that a direct comparison of the frequency with which male-named and female-named storms caused fatalities, caused 1 to 99 fatalities, or caused more than 99 fatalities does not show a consistent pattern, let alone statistically significant differences. A comparison of the average number of fatalities shows that male-named storms generally

Table 6.1 *Fatalities From Atlantic hurricanes and tropical storms*

	Number of storms		Average fatalities	
	Female names	Male names	Female names	Male names
All storms	210	210	36	131
Storms with fatalities	111	103	68	267
Storms with 1 to 99 fatalities	103	88	12	14
Storms with more than 99 fatalities	8	15	792	1,752

Table 6.2 *Average fatalities From Pacific hurricanes and tropical storms*

	Number of storms		Average fatalities	
	Female names	Male names	Female names	Male names
All storms	293	286	4	8
Storms with fatalities	42	46	28	51
Storms with 1 to 99 fatalities	39	42	9	7
Storms with more than 99 fatalities	3	4	271	519

had a higher average number of fatalities, though none of the differences are statistically significant.

The *PNAS* study's assertion that female-named storms are deadlier than male-named storms is not robust, evidently because it relied on the questionable statistical analysis of narrowly defined data from years when all hurricanes had female names.

Why do good journals published flawed research? Some don't have the resources for careful scrutiny. Some like the publicity that comes with provocative articles. Some welcome confirmation of their prejudices.

The Laugher Curve

When Ronald Reagan was elected U.S. President in 1980, the highest income tax bracket was 70 percent. This was a marginal tax rate in that it was levied on income above certain specified levels, with income below those thresholds taxed at lower rates. For single tax payers, the 70 percent tax rate applied to income above $108,300 ($302,000 in 2016 dollars). For married couples, the 70 percent bracket started at $215,400 ($600,000 in 2016 dollars). In practice, wealthy households usually found ways to reduce their tax burden but, still, a 70 percent tax is daunting. Reagan proposed cutting tax rates, a policy that came to be known as *Reaganomics*.

In 1981, I had lunch with an economist ("John") who had started a consulting firm to advise Reagan's economic team. Many people, even traditional Republicans, were concerned that Reagan's proposed tax cuts would increase the government's budget deficit. Reagan's team dismissed such worries and argued that we needed to take a fresh approach—what they called *supply side economics.*

Traditional Keynesian demand-side economists argued that business cycles are caused by changes in aggregate demand. A recession might start when households cut back on spending, causing other people to lose their jobs and reduce their spending, like a snowball rolling downhill, gaining size and momentum along the way. In demand-side models, recessions can be slowed or stopped by tax cuts that give people more money to spend.

Supply-side economists, in contrast, focus on work decisions rather than spending decisions, and emphasize the disincentive effects of high income taxes. If banana taxes are increased, people won't buy as many bananas. If income taxes are increased, people won't work as hard to earn income.

Often, the simplest stories are the most persuasive. Figure 1 shows the Laffer curve, which was reportedly sketched on a restaurant napkin by economist Art Laffer.

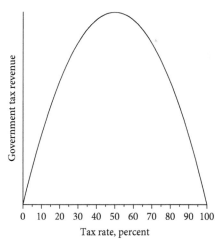

Figure 1 The Laffer curve

The two extremes are clear. At a 0 percent tax rate, no tax revenue would be collected. At a 100 percent tax rate, no one would work and, again, no tax revenue would be collected. In between these extremes, some people will work and some tax revenue will be collected. If tax revenue initially rises as the tax rate is increased from 0 percent, it must also decline at some point in order to get back to zero revenue at a 100 percent tax rate.

Some critics called it the Laugher Curve. There is no reason why the curve should be symmetrical with a peak at a 50 percent tax rate. Nor does the curve need to be smooth with only one peak. The supply-side argument also ignores the reality that most people do not have the luxury of choosing how many hours to work or quitting the jobs they have. It is also conceivable that higher income taxes will force some people to take second jobs in order to make their mortgage payments and pay their other bills.

It is good to consider both demand and supply (how could an economist believe otherwise?), but the supply-side benefits of tax cuts seemed exaggerated by supply-siders. Still, I agreed to have lunch with John in 1981, and it was an unexpected eye-opener.

John's firm was being paid to develop a computer model that would predict an increase in government tax revenue if tax rates were reduced. This model could then be cited by the Reagan administration as evidence that a cut in tax rates would reduce the government's budget deficit.

John knew what his model was supposed to do, but he couldn't figure out how to do it. He tried all sorts of models, but every time he estimated his models with real data, his models predicted that cutting tax rates would reduce tax revenue. He invited me to lunch because he was desperate for suggestions. Maybe I could suggest a new model? Maybe if he only looked at certain historical periods? Maybe if he looked at other countries?

He was not pleased with my suggestion that maybe he should accept the fact that reducing tax rates would reduce tax revenue.

The Reagan administration never publicized John's model, so my guess is that he was never able to torture the data sufficiently to get the conclusion he wanted. Nonetheless, Reagan got the tax cuts he wanted, with the top tax rate reduced from 70 percent to 50 percent, then 38.5 percent, and finally 28 percent when Reagan left office in 1988.

However, as predicted by virtually all professional economists, including many supply-siders, income tax revenue declined.

What is especially interesting about this experience is the Reagan administration's assumption that a prediction of increased tax revenue would be taken more seriously if it came from a computer. There seems to be considerable truth in that assumption. A recent study asked dozens of volunteers to read a paragraph about a fictional medication. Half included a graph; half did not. The graph was just a simple visual display of information included in the paragraph, yet it increased the number of people who were persuaded that the medication would "really reduce illness" from 67.7 percent to 96.6 percent. When things look scientific, they seem more believable. When things come from a computer, they should be believed.

It's not you, it's me

A related gullibility is our willingness to believe that if somebody says something that we don't understand, the fault must be with us. We assume that the person is an expert and that we are not smart enough to understand what this clever person is saying. Similarly, if we don't understand the conclusions given by a computer, the fault must be with us. We are not smart enough to understand what the computer is saying.

A wonderful example of this natural human gullibility involved *Social Text*, an academic journal published by Duke University Press, and touted as "a journal at the forefront of cultural theory." Alan Sokal, a professor of physics at New York University and University College London, submitted a paper with the imposing title, "Transgressing the Boundaries: Towards a Transformative Hermeneutics of Quantum Gravity." The article was essentially unreadable. Here is a sample: "Analogous topological structures arise in quantum gravity, but inasmuch as the manifolds involved are multidimensional rather than two-dimensional, higher homology groups play a role." The journal's humanistic editors were duly impressed and published the article.

It was a prank. Sokal had deliberately written an unintelligible article to see if it would be published, and it was.

The journal's editors were given one of the Ig Nobel Prizes that are awarded annually at a hilarious Harvard ceremony for "achievements that first make people laugh, and then make them think." This Ig Nobel award saluted the editors of *Social Text* for "eagerly publishing research that they could not understand, that the author said was meaningless, and which claimed that reality does not exist."

The editors were not amused. They collectively protested Sokal's hoax as unethical, though one of the editors still "suspected that Sokal's parody was nothing of the sort."

In an interesting follow-up, Robb Willer, a Cornell graduate student, conducted an experiment in which participants were asked to evaluate excerpts from Sokal's unintelligible paper based on the quality of the arguments and intelligibility of the writing. Half were told that the author was a distinguished Harvard professor; the other half were told that the author was a Cornell sophomore. Not surprisingly, the participants gave higher ratings when the author was thought to be a Harvard professor.

Great to good

Jim Collins spent five years studying the 40-year history of 1,435 stocks and identified 11 that outperformed the overall market:

Abbott Laboratories	Kimberly-Clark	Pitney Bowes
Circuit City	Kroger	Walgreens
Fannie Mae	Nucor	Wells Fargo
Gillette	Philip Morris	

These companies were compared to 11 companies in the same industry whose stocks had languished. Collins identified five distinguishing traits, such as Level 5 Leadership (leaders who are personally humble, but professionally driven to make a company great). These are *exactly* the kinds of conclusions that data-mining software might reach after searching for formulas/secrets/recipes for a successful business, a lasting marriage, living to be 100, and so on and so forth. We know that there are sure to be common characteristics, and finding them proves nothing.

In his book, *Good to Great*, summarizing the results, Collins wrote that:

we developed all of the concepts in this book by making empirical deductions directly from the data. We did not begin this project with a theory to test or prove. We sought to build a theory from the ground up, derived directly from the evidence.

Collins thought he was making his study sound unbiased and professional. He didn't just make this stuff up. He went wherever the data took him. In reality, Collins was admitting that he had used Texas Sharpshooter Fallacy #2 and that he was blissfully unaware of the perils of data mining—deriving theories from data.

When we look back in time at any group of companies, the best or the worst, we can always find some common characteristics. Look, every one of those 11 companies selected by Collins has either an *i* or an *r* in its name, and several have both an *i* and an *r*. Is the key for going from good to great to make sure that your company's name has an *i* or *r* in it? Of course not.

To buttress the statistical legitimacy of his theory, Collins cited a professor at the University of Colorado: "What is the probability of finding by chance a group of 11 companies, all of whose members display the primary traits you discovered while the direct comparisons do not possess those traits?" The professor calculated this probability to be less than 1 in 17 million.

This is the Feynman Trap, coincidentally with the same 1-in-17-million probability as in Feynman's license-plate calculation. Finding common characteristics *after* the companies have been selected is not unexpected, or interesting. The interesting question is whether these common characteristics are of any use in predicting which companies will succeed *in the future*. For these 11 companies, the answer is no. Fannie Mae stock went from above $80 a share in 2001 to less than $1 a share in 2008. Circuit City went bankrupt in 2009. The performance of the other nine stocks after the publication of *Good to Great* was distinctly mediocre, with five stocks doing better than the overall stock market, six doing worse.

Identifying successful companies and discovering common characteristics (which is exactly what data-mining software would do) is meaningless because, after the fact, there are *always* successful companies and they *always* have common characteristics.

I did a silly little experiment to prove this. I have several years of grades in a statistics class I teach at Pomona College. If I had access to data on student characteristics (heights, weights, gender, race, movie preferences, music preferences, high school size, number of languages spoken, and so on), I could surely find some distinguishing traits and conclude that these are the secrets for doing well in my statistics class. We could surely think up explanations to fit the correlations: students with low BMI ratings are more active, students from small high schools are more confident. If these same characteristics turned out to be inversely correlated with stats grades, we could invent explanations for that too: student with low BMI ratings spend too much time exercising; students from small high schools were not challenged.

Lacking such data, I used the data I had—student names. The name Gary, for example, has a first-letter G, second-letter A, third-letter R, and fourth-letter Y. I used some data mining software to explore how statistics course grades are related to letter positions. The grades are on a four-point scale: An *A* is 4.0, a *B* 3.0, and so on. The average grade was 3.03, slightly above a *B*.

For first names, the data-mining algorithm found that having a D as the seventh-letter of one's name raised a student's grade by 0.94; a second-letter *D* increased the grade by 0.81 and a fifth-letter *D* increased the grade by 0.79. On the other hand, a fourth-letter *D* dropped the grade by 0.96. The worst thing to have is an eighth-letter *G*, which reduced the grade by 1.85.

For the last name, a fourth-letter *V*, sixth-letter *B*, and fifth-letter *C* all add about a point, while a second-letter *M* reduces the grade by two points.

This is all utter nonsense, although the data-mining program did not know that it is nonsense, because it does not know what grades and names are. If I had done exactly the same experiment using heights, weights, and so on, it might have been tempting to invent explanations for the knowledge discovery. I deliberately chose meaningless characteristics to demonstrate that patterns can always be found, after the fact, that are correlated with the success of companies or students. Finding such characteristics is to be expected and does not demonstrate that they are the keys to success.

Aggressiveness and attractiveness

One dozen married women in their forties watched videos of 20 married men in their twenties playing pickup basketball games. The women rated each man on his aggressiveness and physical attractiveness, and the ratings were averaged to give a rank order, with 1 being the highest rated and 20 the lowest.

The scatter plot of the rankings in Figure 2 shows that men who were ranked higher in aggressiveness were generally ranked lower in attractiveness. The correlation is an impressive -0.93. Despite the popular belief that girls are attracted to "bad boys," this study indicates that women in their forties are not attracted to aggressive men, perhaps because they have learned that dangerous and reckless behavior can have serious consequences. They are attracted to men who are stable and can control themselves.

I can see the internet story now. Guys, behave yourselves and put away those testosterone pills.

Are you persuaded? The results, statistical analysis, and graph all came from a computer, so they must be correct. Well, the fact that the calculations are correct mathematically doesn't necessarily mean that the conclusions are persuasive.

What if I were to tell you that the relationship between aggression and attractiveness shown in Figure 2 was found by rummaging through a large data set that contained 1,000 characteristics for these 20 guys? Most of the characteristics were largely unrelated, but there were some striking statistical relationships—and the most striking was this negative relationship between aggression and attractiveness.

True, the professor who did this study was not looking for this specific relationship, but there clearly is one. True, if the relationship had turned out to be positive, the professor would have reached a different

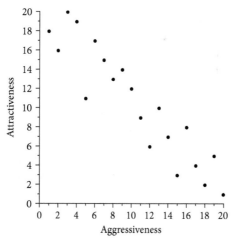

Figure 2 Nice guys finish first

conclusion—that women are attracted to aggressive men. But, still, and this is the most important point, it is indisputable that the professor found a very strong empirical relationship.

Now, a cynic might argue that the professor looked at correlations among a thousand variables. Even if the data were just meaningless noise, the professor was bound to find some strong statistical relationships by luck alone.

The professor might be incensed and point out that these are not random numbers. They are real characteristics and they reveal a real relationship between aggression and attractiveness. No one can dispute the strength of that relationship. Because the statistical relationship is so clear, the professor must have discovered an important truth. That is what data mining and knowledge discovery are all about.

Okay, time for a confession: There were no videos or rankings. I am the professor and I used a computer random number generator to scramble the numbers 1 to 20 randomly 1,000 times. I used a data mining program to ransack these data looking for "unexpected" relationships. Figure 2 shows the data for random rankings 462 and 594—which I labeled aggression and attractiveness after I discovered them. And that is precisely the point. Ransacking large data bases for statistical relationships can discover statistically impressive relationships even if the data are just random numbers. Discovering such relationships does not prove anything beyond the fact that the data were ransacked.

Data mining may not be knowledge discovery, but noise discovery.

The Dartmouth salmon study

A standard neuroscience experiment involves a volunteer in an MRI machine who is shown various images and asked questions about the images. Functional magnetic resonance imaging (fMRI) measures the magnetic disruption that happens as oxygenated and deoxygenated blood flows through the brain. After the test, the researchers look at up to 130,000 voxels (3-D data) to see which parts of the brain were stimulated by the images and questions.

The fMRI measurements are noisy, picking up magnetic signals from the environment and from variations in the density of fatty tissue in different parts of the brain. Sometimes the voxels miss brain activity; sometimes they suggest activity where there is none.

A Dartmouth graduate student named Craig Bennett did an unusual experiment when he used an MRI machine to study the brain activity of a salmon as it was shown photographs and asked questions. A sophisticated statistical analysis found some clear patterns.

The most interesting thing about the study was not that a salmon was studied, but that the salmon was dead. Yep, Bennett put a dead salmon he purchased at a local market into the MRI machine, showed it photographs, and asked it questions. With so many voxels, some random noise was recorded, which might be interpreted as the salmon's reaction to the photos and questions. Except that the salmon was dead.

This dead-salmon study got more press than most fMRI studies, even winning an Ig Nobel Prize.

This study is wonderfully analogous to someone data-mining Big Data looking for patterns, except that Big Data contain far more data and can yield far more ludicrous relationships.

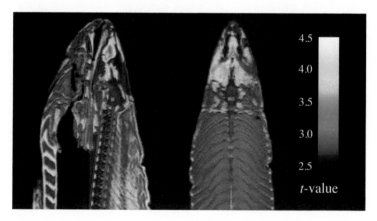

Quack, quack

Ever hopeful, people scour data looking for ways to beat the stock market and win the lottery, and come up will ridiculous theories involving the Super Bowl or having a friend named Mary buy their lottery tickets. Don't fall for such nonsense.

We can always find patterns—even in randomly generated data—if we look hard enough. No matter how stunning the pattern, we still need a plausible theory to explain the pattern. Otherwise, we have nothing more than coincidence. Several more examples are in my book *Standard Deviations: Flawed Assumptions, Tortured Data, and Other Ways to Lie With Statistics*, including claims that Asian-Americans are susceptible to heart attacks on the fourth day of every month and that people can postpone their deaths in order to celebrate important occasions.

Being human, we can recognize implausible claims and remember the dead salmon study, computers cannot.

Yes, human researchers data-mine too much, and this has created a replication crisis. Serious researchers are now trying to find effective ways to curtail data mining. One proposal is that journals should not insist on statistically significant results, since this is the hurdle that lures people into the data-mining trap.

AI algorithms that mine Big Data are a giant step in the wrong direction—a wrong step that can turn the replication crisis into a replication catastrophe. Computers can never understand in any meaningful way the fundamental truth that models that make sense are more useful than models that merely fit the data well. What AI algorithms do understand is how to torture data.

CHAPTER 7

The kitchen sink

Back in the 1980s, I talked to an economics professor who made forecasts for a large bank based on simple correlations like the one in Figure 1. If he wanted to forecast consumer spending, he made a scatter plot of income and spending and used a transparent ruler to draw a line that seemed to fit the data. If the scatter looked like Figure 1, then when income went up, he predicted that spending would go up.

The problem with his simple scatter plots is that the world is not simple. Income affects spending, but so does wealth. What if this professor happened to draw his scatter plot using data from a historical period in which income rose (increasing spending) but the stock market crashed (reducing spending) and the wealth effect was more powerful than the income effect, so that spending declined, as in Figure 2? The professor's scatter plot of spending and income will indicate that an increase in income reduces spending. Then, when he tries to forecast spending for a period when income and wealth both increase, his prediction of a decline in spending will be disastrously wrong.

Multiple regression to the rescue.

Multiple regression models have multiple explanatory variables. For example, a model of consumer spending might be:

$$C = a + bY + cW$$

where C is consumer spending, Y is household income, and W is wealth.

The order in which the explanatory variables are listed does not matter. What does matter is which variables are included in the model and which are left out. A large part of the art of regression analysis is choosing

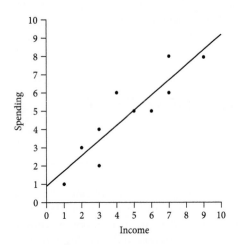

Figure 1 A positive correlation

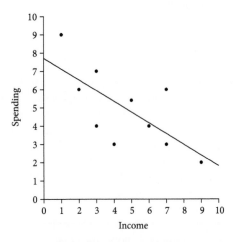

Figure 2 A negative correlation

explanatory variables that are important and ignoring those that are unimportant.

The coefficient b measures the effect on spending of an increase in income, holding wealth constant, and c measures the effect on spending of an increase in wealth, holding income constant. The math for estimating these coefficients is complicated but the principle is simple: choose the estimates that give the best predictions of consumer spending for the data used to estimate the model.

In Chapter 4, we saw that spurious correlations can appear when we compare variables like spending, income, and wealth that all tend to increase over time. To make sure we are not being misled by such spurious correlations, I looked at the annual percentage changes in inflation-adjusted spending, income, and wealth.

I used some statistical software to calculate the regression line for annual U.S. data:

$$C = 0.62 + 0.73Y + 0.09W$$

Holding wealth constant, a 1 percent increase in income is predicted to increase spending by 0.73 percent. Holding income constant, a 1 percent increase in wealth is predicted to increase spending by 0.09 percent. Figure 3 compares the predicted and actual percent changes in spending. The correlation is an impressive 0.82.

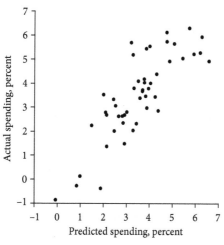

Figure 3 Predicted and actual percent increases in U. S. household spending

The coefficient of wealth may seem small, but changes in wealth are often quite large. There have been several years when wealth rose or fell by more than 10 percent, which our model predicts will reduce consumer spending by 0.9 percent—which can be the difference between economic expansion and recession.

Multiple regression models are extremely powerful—much more powerful than simple correlations—because they take into account the importance of multiple explanatory variables. This is why they are one of the most important—if not the most important—statistical tools.

However, when used for data mining, they are highly susceptible to abuse.

Predicting presidential elections

In my statistics classes, I ask students to name some factors that they think determine the outcomes of presidential elections. They mention the economy, the candidates' personalities, whether the nation is at war, and other factors. I write their suggestions on the whiteboard, and then I show them my model.

For more than 100 years, U.S. presidential elections have generally been two-party races between the Democratic and Republican candidates. The incumbent party candidate is either the President or a nominee from the President's party. Barrack Obama was the incumbent candidate when he ran for a second term in 2012. Hillary Clinton was the incumbent party's candidate when she campaigned in 2016 to replace Obama at the end of his two terms.

The incumbent party has many advantages, including easy access to the media and fund raising. The incumbent party also appeals to people who want stability and are satisfied with the way things are. On the other hand, voters who are dissatisfied with the economy, wars, or other issues may vote for the challenger—the change candidate. It has been estimated that, on balance, the incumbent candidate has a 4 to 6 percent advantage over the challenger, though the actual outcome obviously depends on the specific candidates and historical circumstances.

If I were to tell my students that I could predict the incumbent party's share of the two-party vote based solely on whether the incumbent party's nominee was President, they would think I was joking. And with good reason. We all know of cases where the incumbent president did well (Ronald Reagan with 59 percent of the vote) and cases where the incumbent did poorly (Jimmy Carter with 44 percent of the vote).

But what if I were to also take into account whether the candidates had been governors, senators, and so on? I show my students the following multiple regression model that I estimated using the outcomes of the last ten presidential elections (1980–2016):

$$i\% = 78.31 - 7.35iP - 13.07iV + 7.93cV - 27.20iS$$
$$+ 14.75cS - 34.46iG + 8.20cG - 19.54iR + 3.49cR$$

The variables are:

$i\%$ = Percent of major party vote received by the incumbent party's candidate

iP = 1 if the incumbent party candidate is President, 0 otherwise

iV = 1 if the incumbent party candidate has been a U.S. Vice President, 0 otherwise

cV = 1 if the challenger party candidate has been a U.S. Vice President, 0 otherwise

iS = 1 if the incumbent party candidate has been a U.S. Senator, 0 otherwise

cS = 1 if the challenger party candidate has been a U.S. Senator, 0 otherwise

iG = 1 if the incumbent party candidate has been a state governor, 0 otherwise

cG = 1 if the challenger party candidate has been a state governor, 0 otherwise

iR = 1 if the incumbent party candidate has been a U.S. Representative, 0 otherwise

cR = 1 if the challenger party candidate has been a U.S. Representative, 0 otherwise

I didn't look at the economy, the candidates' personalities, or any of the other factors that you and my students think are important. I selected some vaguely relevant factors and got a perfect relationship, in that my equation predicts the outcome of all ten presidential elections perfectly. For example, my model's prediction of Hillary Clinton's share of the 2016 two-party vote is exactly equal to the actual 51.11 percent share she received.

When my students see that my model fits the data perfectly, they are tempted to think that I have found the holy grail for predicting presidential elections. My model does not include any of the factors that they think matter, but it sort of makes sense in that it uses explanatory variables that are related to the backgrounds of the candidates running for President. Most importantly, my model fits the data perfectly, so it must be correct and they must be wrong.

Then I show them another model that fits these ten presidential elections 1980–2016 perfectly:

$$i\% = 84.79 - 1.62T1 - 0.30T2 - 0.04T3 - 0.54T4$$
$$+ 2.94T5 - 0.39T6 + 0.60T7 + 0.14T8 - 1.05T9$$

These nine explanatory variables are the high temperatures on election day in nine cities in states with few electoral votes:

T1 = high temperature in Bozeman, Montana

T2 = high temperature in Broken Bow, Nebraska

T3 = high temperature in Burlington, Vermont

T4 = high temperature in Caribou, Maine

T5 = high temperature in Cody, Wyoming

T6 = high temperature in Dover, Delaware

T7 = high temperature in Elkins, West Virginia

T8 = high temperature in Fargo, North Dakota

T9 = high temperature in Pocatello, Idaho

I picked these cities because I liked the names and I could find daily weather data back to 1940.

Now my students are puzzled. Many are suspicious. Am I making things up? How could the temperature in Bozeman or Broken Bow have any meaningful effect on presidential elections? Why is the vote for the incumbent party's candidate affected negatively by warm weather in Bozeman, and affected positively by warm weather in Cody? There is no logical explanation, yet the model fits the data perfectly.

Maybe presidential elections are affected by the weather. Maybe this is one of those unexpected relationships uncovered by ransacking data. Maybe I stumbled on a knowledge discovery that demonstrates the power of data mining. Are you tempted to believe this?

So, I make it even more preposterous. I estimated a third model that fits the presidential election data for 1980–2016 perfectly:

$$i\% = 33.73 - 0.01R1 + 0.26R2 + 0.21R3 + 0.20R4$$
$$- 0.01R5 + 0.19R6 + 0.01R7 - 0.33R8 - 0.18R9$$

This time, the explanatory variables really are random. I used a computer software program to generate random two-digit numbers that have nothing at all to do with the real world, let alone what is going on in the United States during presidential election years. Yet, the model fits the data great.

Despite my students' suspicions, I did not make up any of this, but I did have a secret.

The secret sauce

Suppose that I wanted to explain why stock prices were about 10 percent higher at the end of 2016 than at the end of 2015, and I claimed that it was all about the weather—specifically, the weather in Porterville, California, a small town in California's Central Valley where my father grew up. You would think I was nuts, and I would be if I made that claim seriously. But hear me out.

Figure 4 shows a scatter plot of data on the last day of 2015 and 2016 for the S&P 500 index of stock prices and the low temperature in Porterville. There is an absolutely perfect relationship. The correlation between the two variables is 1. Stock prices can be predicted perfectly by tracking temperatures in my dad's hometown. Who knew?

The trick (of course, there is a trick) is that there is always a perfect linear relationship between any two points in a scatter plot. I could have chosen the number of new-born babies named Claire in 1974 and 1997 or the number of wins by the San Antonio Spurs basketball team in 2012

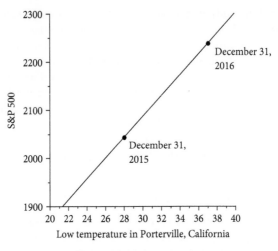

Figure 4 Predicting stock prices

and 2015. There would have been a perfect linear relationship between these numbers and the S&P 500, because any two points always lie on the straight line connecting them.

Yet, the fitted relationship is completely useless. Anyone who tries to predict stock prices based on temperatures in Porterville is destined for failure.

I used two data points on a two-dimensional graph to illustrate the folly that applies to more complicated models using more data. Figure 4 uses one explanatory variable (Porterville temperatures) to give a perfect fit to two observations. If there were three observations, two explanatory variables would give a perfect fit. If there were ten observations, nine explanatory variables would give a perfect fit.

That's how I came up with three increasingly preposterous models for predicting ten presidential elections: I used nine explanatory variables. Period. There is nothing special about these nine explanatory variables. Any nine would do. The important thing is that I used nine variables to predict ten elections.

This is an extreme example of what is called *over-fitting* the data. In any empirical model, I can improve the model's explanatory power by adding more and more explanatory variables—in extreme cases, to the point where the fit is perfect. It hardly matters whether the variables make sense or not.

This is also known as the *kitchen-sink* approach to modeling: throw every explanatory variable but the kitchen sink at the model. The inescapable problem is that even though the model may fit the original data very well, it is useless for predictions using new data. Porterville weather won't predict stock prices accurately, except by chance. My nine-variable presidential election model won't predict other presidential election outcomes accurately, except by chance.

We can illustrate my presidential-election model's flaws by looking back at the ten presidential elections before 1980. Figure 5 shows that the first model, using incumbency/challenger data, fit to the ten elections between 1980 and 2016, is perfect for these ten elections but awful for the ten preceding elections. The model predicts that Richard Nixon would lose the 1972 election by a landslide, getting only 29 percent of the popular vote. In fact he won by a landslide, with 62 percent of the vote. Nixon won every state but Massachusetts, leading some Bay Staters to display bumper stickers that said simply, "Don't blame us."

The predictions for the 1956 and 1964 elections are even worse, predicting that Dwight Eisenhower would get a near-impossibly high 79 percent of the vote in 1956 (he got 58 percent) and predicting that Lyndon Johnson

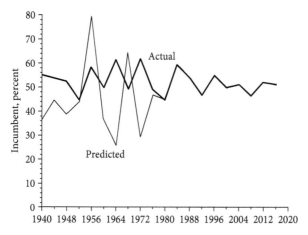

Figure 5 Predicting Presidential elections by overfitting the 1980–2016 data

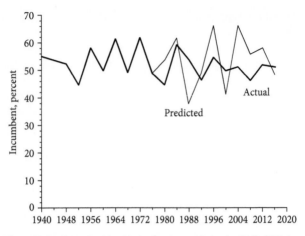

Figure 6 Predicting Presidential elections by overfitting the 1940–1976 data

would get a near-impossibly low 26 percent of the vote in 1964 (he got 61 percent).

I overfit the ten most recent elections and then tried (and failed) to predict earlier elections. I could just as well have estimated the coefficients by overfitting the ten earlier elections and then using this model to predict recent elections. Figure 6 shows that my revised model fits the ten presidential elections between 1940 and 1976 perfectly, but the predictions are lousy for the ten subsequent elections.

The same is true of my other two models. For example, Figure 7 shows that, yes, the temperature model fits the data used to estimate the model perfectly, but does terribly predicting election outcomes in other years. In fact, the model predicts that in the 1940 election, Franklin Roosevelt gets –11 percent (yes, negative 11 percent) of the vote instead of the 55 percent he actually received.

Admittedly, using nine explanatory variables to predict ten elections is an extreme case. I did it to demonstrate the general principle that adding explanatory variables to regression models gives a better fit even if the added variables are nonsensical.

Here, in the presidential-election weather model, we don't need all nine explanatory variables to get a good fit. The correlation between the weather model's predictions and the actual vote is 0.94 even with only five

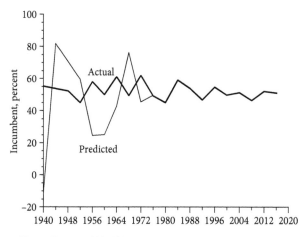

Figure 7 Using 1980–2016 weather to predict Presidential elections

explanatory variables (the weather in Burlington, Cody, Dover, Elkins and Pocatello):

$$i\% = 72.75 - 0.38T3 + 0.59T5 + 0.40T6 - 0.38T7 - 0.65T9$$

You should not be surprised by Figure 8, which shows that this five-variable weather model does a great job fitting the data for 1980–2016 and a terrible job fitting the data for 1940–1976:

We can get a good fit with even fewer explanatory variables. A model with four cities (Burlington, Cody, Elkins, and Pocatello) has a 0.86 correlation with the election outcomes in 1980–2016 and a model with three cities (Cody, Elkins, and Pocatello) has a 0.79 correlation.

The same is true if we fit the model to the 1940–1976 data. A model with four cities (Broken Bow, Dover, Elkins, and Fargo) has a 0.89 correlation with the election outcomes in 1940–1976; and a model with three cities (Broken Bow, Elkins, and Fargo) has a 0.86 correlation.

How did I choose these particular cities? I had daily data on the high and low temperatures in 25 cities and I used data-mining software to consider all possible combinations of these 50 variables, and identify the ones that fit the presidential election results the closest.

The cities that give the best fit in 1980–2016 are mostly different from the cities that give the best fit in 1940–1976 because there is no logical

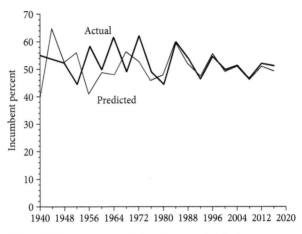

Figure 8 Using temperatures in five cities to predict election outcomes.

basis for the model. These are essentially random data that I used to find spurious correlations. Any half-decent data mining program would come up with the same nonsense—and not know that it is nonsense.

The random-variable model also fits the data really well even if the number of explanatory variables is reduced to five (0.97 correlation), four (0.95 correlation), or three (0.89 correlation). Everything is pretty much the same as with the weather model, including the fact that it is utterly useless in predicting election outcomes in years that were not used to estimate the model.

The conclusion is inescapable. Data-mining expeditions can easily discover models with multiple explanatory variables that can fit data surprisingly well even if the explanatory variables have nothing whatsoever to do with variable being predicted. Unfortunately, data-mining programs cannot assess whether the models make sense because, to computer software, numbers are just numbers.

How can we tell if a discovered pattern is real or spurious? Only by using human knowledge of the variables to judge whether there is a logical basis for the unearthed patterns.

I've belabored this point because, over and over again, I have spoken with intelligent and well-intentioned people who do not fully appreciate how easy it is to find coincidental patterns and relationships. That includes most data miners I have spoken with. Many are vaguely aware of the possibility of spurious correlations, but nonetheless believe that statistical evidence of patterns and relationships is sufficient proof that they are real.

In 2017 Greg Ip, the chief economics commentator for the *Wall Street Journal*, interviewed the co-founder of a company that develops AI applications for businesses. Ip paraphrases the co-founder's argument:

If you took statistics in college, you learned how to use inputs to predict an output, such as predicting mortality based on body mass, cholesterol and smoking. You added or removed inputs to improve the "fit" of the model.

Machine learning is statistics on steroids: It uses powerful algorithms and computers to analyze far more inputs, such as the millions of pixels in a digital picture, and not just numbers but images and sounds. It turns combinations of variables into yet more variables, until it maximizes its success on questions such as "is this a picture of a dog" or at tasks such as "persuade the viewer to click on this link."

No! What statistics students *should* learn in college is that it is perilous to add and remove inputs simply to improve the fit. The same is true of machine learning. Rummaging through numbers, images, and sounds for the best fit is mindless data mining—and the more inputs considered, the more likely it is that the selected variables will be spurious.

The fundamental problem with data mining is that it is very good at finding models that fit the data, but totally useless in gauging whether the models are ludicrous. Statistical correlations are a poor substitute for expertise.

The best way to build models of the real world is to start with theories that are appealing (like the state of the economy affects presidential elections) and then test such models. Models that make sense can be used to make useful predictions outside the data used to estimate the models. Data mining works backwards, with no underlying theory, and therefore cannot distinguish between models that are reasonable and those that are ridiculous, which is why their predictions for fresh data are unreliable.

Nonlinear models

In addition to overfitting the data by sifting through a kitchen sink of explanatory variables, data-mining algorithms can overfit the data by trying a wide variety of nonlinear models.

Figure 9 shows a simple scatter plot using hypothetical data. A straight line (a linear model) does not fit all three observations perfectly, but gives a pretty good fit that might be useful in predicting Y if there really is a causal relationship between X and Y.

Figure 10 shows that a nonlinear model gives a perfect fit to these three data points. Is the nonlinear model in Figure 10 therefore an improvement over the linear model in Figure 9? Not necessarily, and there is no reasonable way for a data-mining algorithm to decide.

The model in Figure 9 implies that, as X increases, Y will increase, too, by a constant amount. The model in Figure 10 implies that as X increases, the increases in Y become smaller and then turn negative. After X goes above 7, Y is negative.

Which of these models is more useful for predicting Y for values of X that were not used to fit the model? It depends. If X is household income and Y is spending, it is plausible that spending keeps increasing by a roughly constant amount as income increases, as in Figure 9. It is not plausible that an increase in income will, at some point, cause spending to fall, as in Figure 10, and, farther along, cause spending to be negative.

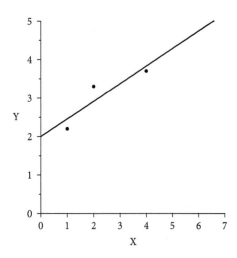

Figure 9 A linear model does not fit these three observations perfectly

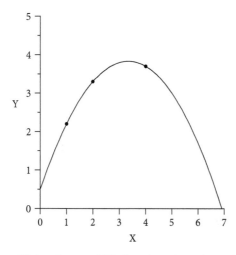

Figure 10 A nonlinear model fits these three observations perfectly

Suppose, on the other hand, that X is the amount of nitrogen fed to tomato plants and Y is the subsequent plant growth. Now, Figure 9 seems implausible with its assumption that every extra dollop of nitrogen has the same positive effect on plant growth. Figure 10 may be more plausible. As more and more nitrogen is supplied, the effects on growth may diminish and there may come a point where additional nitrogen stunts growth. There may even come a point where the plant is killed by too much nitrogen.

How can a data-mining algorithm decide whether the linear model in Figure 9 or the nonlinear model in Figure 10 is a better representation of the reality that is being modeled? Certainly not by seeing which model fits the data better! We can only choose between these models, or other models, by using expert opinion (i.e., humans) to assess which model is more realistic.

Figure 11 shows an even more extreme example. If there is a logical explanation, it seems perfectly reasonable to fit a straight line to these data and to interpret the variation in the points about the line as inevitable fluctuations due to factors that are not in the model. Unless something changes drastically, the straight line should make reasonably accurate predictions.

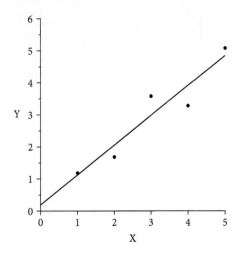

Figure 11 A reasonable linear model

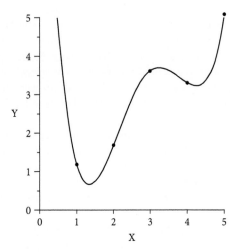

Figure 12 An unreasonable nonlinear model

Figure 12 shows what happens when a data-mining algorithm runs amok by selecting an overly complicated nonlinear model to fit the data perfectly. Despite the perfect fit to the original data, the nonlinear model's predictions for new values of X are bound to be wildly inaccurate and sometimes bizarre.

The issue, as always, is that data mining algorithms look for patterns (and are very good at finding them), but have no way of assessing the patterns they find. Words like *spending*, *income*, and *wealth* are just combinations of letters, like Nigel Richards playing Scrabble in a language he does not understand. Computer algorithm cannot tell which explanatory variables belong in a model. Computer algorithms cannot tell whether a linear or nonlinear model is more sensible. These decisions require human intelligence.

CHAPTER 8

Old wine in new bottles

The fashion industry is subject to recurring cycles of popularity that are regular enough to be dubbed the "20-year rule." Activewear clothing that is suitable for the gym and the street was fashionable in the 1990s and, 20 years later, in the 2010s.

Intellectually, the British economist Dennis Robertson once wrote, "Now, as I have often pointed out to my students, some of whom have been brought up in sporting circles, high-brow opinion is like a hunted hare; if you stand in the same place, or nearly the same place, it can be relied upon to come round to you in a circle."

In the same way, today's data miners have rediscovered several statistical tools that were once fashionable. These tools have been given new life because they are mathematically complex, indeed beautifully complex, and many data miners are easily seduced by mathematical beauty. Too few think about whether the underlying assumptions make sense and if the conclusions are reasonable.

Stepwise regression

Consider data mining with multiple regression models. Rummaging through a large data base looking for the combination of explanatory variables that gives the best fit can be daunting. With 100 variables to choose from, there are more than 17 trillion possible combinations of 10 explanatory variables. With 1,000 possible explanatory variables, there are nearly a trillion trillion possible combinations of 10 explanatory variables.

With 1 million possible explanatory variables, the number of 10-variable combinations grows to nearly a million trillion trillion trillion trillion (if we were to write it out, there would be 54 zeros).

Stepwise regression was born back when computers were much slower than today, but it has become a popular data-mining tool because it is less computationally demanding than a full search over all possible combinations of explanatory variables but, it is hoped, will still give a reasonable approximation to the results of a full search.

The *stepwise* label comes from the fact that the calculations go through a number of steps, considering potential explanatory variables one by one. There are three main stepwise procedures.

A forward-selection rule starts with the one explanatory variable that has the highest correlation with the variable being predicted. Then the procedure adds a second variable, the variable that improves the fit the most. Then the best third variable, and so on until the increase in the fit does not meet some pre-specified minimum. A backward-elimination rule starts with all possible explanatory variables and then discards variables, one by one, in each case throwing out the variable that has the smallest effect on the fit.

Another popular stepwise procedure is bi-directional, a combination of forward selection and backward elimination. As with forward selection, the data mining starts with the one explanatory variable with the highest correlation with the variable being predicted and adds variables one by one. The wrinkle is that, at every step, the procedure also considers the statistical consequences of dropping variables that were previously included. So, a variable might be added in Step 2, dropped in Step 5, and added again in Step 9.

What do all these stepwise procedures have in common? They are all automated rules that only consider statistical correlations, with no regard whatsoever for whether it makes sense to include or exclude the variables. It is data mining on speed dial.

Others recommend adding or subtracting variables from a regression model based on whether the coefficients are "large" or "small," whatever that means. No, I am not making this up. How could a software program compare the coefficients of an incumbent candidate being a U.S. Senator and the temperature in Broken Bow, Nebraska? In addition, variables should be included in a model because, logically, they belong in the model, not based on the size of their estimated coefficients.

I did some experiments with models that had 100 values of five *true* variables that really did determine what was being predicted and several *fake* variables that were created by a random number generator, but might be coincidentally correlated with the predicted variable. When I applied a stepwise regression procedure, I found that, if 100 variables were considered, 5 true and 95 fake, the variables selected by the stepwise procedure were more likely to be fake variables than real variables. When I increased the number of variables being considered to 200 and then 250, the probability that an included variable was a true variable fell to 15 percent and 7 percent, respectively. Most of the variables that ended up being included in the stepwise model were there because of spurious correlations which are of no help and may hinder predictions using fresh data.

Stepwise regression is intended to help researchers deal with Big Data but, ironically, the bigger the data, the more likely stepwise regression is to be misleading. Big Data is the problem, and stepwise regression is not the solution.

Ridge regression

I saw a presentation in 2013 where ridge regression was called *modern regression*. Ridge regression was developed in the 1970s and discredited in the 1980s, but has re-emerged in the data-miners' tool box with the preposterous label, modern regression. Variations are known as Tikhonov regularization, the Phillips–Twomey method, linear regularization, constrained linear inversion, and other impressive titles. Despite the fancy new labels, ridge regression is not new, and it is no more compelling now than it was in the past. It is old wine in new bottles.

Consider this model of consumer spending developed by Nobel laureate Milton Friedman:

$$C = a + bY + cP$$

where C is spending, Y is current income, and P is permanent income. The idea is that households don't live hand-to-mouth, basing their spending decisions solely on how much income they are currently earning. They also take into account their average or "permanent" income.

People who experience a temporary drop in income usually try to maintain their lifestyle. They must keep paying their mortgage or rent or they will be evicted. They don't want to pull their children out of extracurricular activities. They don't want to downsize their lifestyle unless forced to do so.

On the other hand, people who experience a temporary increase in income may not rush to upgrade their lifestyle, buying a new house and putting their children in private schools, because they will have to retrench when their income returns to normal.

Following this line of reasoning, Friedman argued that a household's spending is affected not only by current income, but also by its perceived long-term average income, which he called permanent income.

One of the most appealing features of multiple regression is its ability to estimate the effect of one explanatory variable, holding the other explanatory variables constant. In our spending model, $C = a + bY + cP$, the coefficient b predicts the effect on spending of an increase in current income, holding permanent income constant. The coefficient c predicts the effect on spending of an increase in permanent income, holding current income constant.

One potential problem is that if current income and permanent income are highly correlated, we won't be able to obtain accurate estimates of their individual effects on spending. For an extreme example, suppose that income and permanent income both increase by $3,000 every year. If spending increases by $2,000, is it because current income went up $3,000 or because permanent income went up $3,000? There is no way of telling. In less extreme cases, where current and permanent income are highly correlated, but not perfectly correlated, we can estimate the separate effects, but our estimates may be very imprecise.

Ridge regression attempts to solve this impasse by adding random noise to the explanatory variables. For every observed value of current income and permanent income, add or subtract a number determined by a random number generator. I am not joking, but I am laughing.

Ridge regression is described by complex mathematical equations, but it is exactly equivalent to adding random noise to the explanatory variables. This random noise makes the explanatory variables less correlated, creating the illusion of more precise estimates. The one-liner for adding self-inflicted errors to the data is, "Use less precise data to get more precise estimates."

How could a data miner believe that estimates are improved by using less accurate data? And yet they do. Data miners are too focused on the math to stop to think about the fact that the variables in their models are not just mathematical symbols. These are real variables that are being used to predict another real variable. Anyone who thought about Friedman's model would surely recognize that measuring income less accurately is not going to improve our spending predictions.

There is a second big problem. Friedman's permanent-income model of consumer spending can also be written as a model of saving, which is income minus spending. Instead of saying that a $1,000 increase in income will increase spending by $700, we could say saving will increase by $300. These are equivalent ways of saying the same thing.

If spending and saving equations are estimated by multiple regression, the results are equivalent. Not so with ridge regression! The results depend on which equation is estimated and there is no basis for choosing.

One of the fuzzy parts of ridge regression is that many practitioners scrutinize something called the ridge trace in order to determine how much random noise to add to the data. I sent some data for four equivalent representations of Friedman's model to a ridge enthusiast. I did not tell him that the data were for equivalent equations. The flimsy foundation of ridge regression was confirmed in my mind by the fact that he did not ask me anything about the data he was analyzing. They were just numbers to be manipulated. He was just like a computer. Numbers are numbers. Who knows or cares what they represent? He estimated the models and returned four contradictory sets of ridge estimates.

It is not just ridge regression. Too many statistical procedures used by data miners are essentially black box. They put numbers in the black box and get out the results. They do not consider what the data measure or what they are to be used for.

Data reduction

Chapter 7 discussed the problem of overfitting data in multiple regression models. When there are ten observations (for example, ten presidential elections), any nine explanatory variables will give a perfect fit— suggesting that the model is perfect when it is not.

If there are more than nine explanatory variables, there are an infinite number of ways to get a perfect fit. The general principle is that when the number of explanatory variables is as large as, or larger then, the number of observations, there are an infinite number of ways to fit the data perfectly. This can clearly be a problem for Big Data when there are millions of variables to be data-mined.

There is also the simple issue of practicality when dealing with a large number of explanatory variables. Suppose that we have 100 potential explanatory variables and want to data-mine these variables looking for the 5 explanatory variables that give the best fit for predicting the S&P 500. We would have to estimate more than 75 million regression equations to find the best fit. To find the 10 best explanatory variables, we would have to estimate more than 17 trillion multiple regression models. And, in practice, there are a *lot* more than 100 potential explanatory variables.

What's a data miner to do when there are so many explanatory variables and so little time? Two statistical tools—principal components and factor analysis—have been around for more than 100 years and have been rediscovered by data miners as data-reduction tools, ways to reduce the effective number of explanatory variables.

Principal components and factor analysis are similar to ridge regression in that they are based on the statistical properties of the variables, with no concern for what the numbers represent. Because principal components and factor analysis are so similar, I will just discuss principal components.

Suppose we are trying to predict the chances that a person will be involved in a car accident and we consider these five variables:

F = gender, 1 if female or 0 if male
Y = age if under 30, 0 otherwise
O = age if over 60, 0 otherwise
T = number of traffic tickets
D = average number of miles driven each year

If these variables are correlated, then it is somewhat redundant to look at all five. Instead, we might use principal components to determine two weighted averages:

$$C1 = 0.30F + 0.01Y + 0.26O + 0.05T + 0.31D$$
$$C2 = 0.26F + 0.19Y + 0.24O + 0.12T + 0.14D$$

Then we can estimate a multiple regression equation predicting the chances of an accident by using these two principal components as explanatory variables in place of the original five variables.

Data miners realized that this variable reduction solves the problem of too many variables. Instead of estimating a model with 1,000 explanatory variables, they can estimate a model with, say, 5 or 10 principal components. However, and this is a big however, the principal components that are most useful for reducing the number of explanatory variables may be less useful for predicting something, like whether a person will be involved in a car accident.

In our example, the principal-component weights used in $C1$ and $C2$ are based on correlations among the five original variables, and tell us nothing about the relative importance of gender, the two age variables, tickets, and miles for being accident-prone. Even though gender is given a higher weight than the number of tickets for the purpose of summarizing these five variables, tickets may be much more important for predicting traffic accidents. A goal of summarizing data is very different from a goal of using the data to make predictions.

Even worse, in a data-mining adventure, some of the original variables may have nothing at all to do with being accident-prone; yet, they might be included in the principal components. Remember, the principal components are chosen based on the statistical relationships among the explanatory variables, with no consideration whatsoever of what these components will be used to predict.

For example, a person's birth month or favorite candy might end up being included among the principal components used to predict whether someone will be involved in a car accident. More generally, if the principal components are based on a data-mining exploration of hundreds or thousands of variables, it is virtually certain that the components will include nonsense. Once again, Big Data is the problem, and principal components is not the solution.

Another issue is that it is generally difficult, if not impossible, to interpret transformed variables like $C1$ and $C2$ and, so, it is problematic to assess whether the estimated regression coefficients are sensible. Do we expect accident frequency to be related positively or negatively to $C1$ and $C2$? Are the estimated coefficients reasonable or implausible? Computer software never knows and, with principal components, humans do not know either. No one knows.

Yet another problem is that, as with ridge regression, unimportant changes in the explanatory variables affect the implicit estimates of the model's coefficients. Suppose that instead of:

$$F = \text{gender, 1 if female or 0 if male}$$

we use this equivalent variable:

$$M = \text{gender, 1 if male or 0 if female}$$

This switch is unimportant and won't affect the multiple regression estimates or the predicted accident frequency of, say, a 47-year-old female who has been ticketed twice and drives an average of 10,000 miles a year.

Yet, this simple switch will change the principal components and give different predicted accident frequencies. As with ridge regression, this is a fatal flaw because there is no logical way to choose among equivalent representations.

Neural network algorithms

Neural network models were first constructed in the 1940s, fell out of favor, and are again fashionable—old wine in new bottles. The label *neural networks* suggests that these algorithms replicate the neural networks in human brains that connect electrically excitable cells called *neurons*. They don't. We do not understand how human brains use neurons to receive, store, and process information, so we cannot conceivably mimic them with computers.

Neural network algorithms are a statistical procedure for classifying inputs (such as numbers, words, pixels, or sounds) so that these data can mapped into outputs. Although the math is intimidating, neural networks are similar to other data-mining algorithms. Neural networks essentially construct weighted linear combinations of input variables, much like principal components, and then use these combinations, much like multiple regression, to estimate nonlinear statistical models that best fit the data being predicted.

The process of estimating the neural-network weights is advertised as *machine learning* from *training data*, suggesting that neural networks function like the human mind; but neural network calculations are not the way humans learn from experience. Instead, neural networks are similar to the way coefficients are estimated in regression models, by finding the values for which the model's predictions are closest to the observed values, with no consideration of what is being modeled.

Neural network algorithms are useful—for example, for language translation and visual recognition—and they will become more useful, but it is misleading to think that they replicate the way humans think. Consider this. In a standard regression model, there are data, like the simple scatter plot in Figure 1, and the regression algorithm fits a line to the data, as in Figure 2. We could call the data *training data* and we could call the fitted line *machine learning*, but that would be misleading, even pretentious, because fitting a line to data is not comparable to how humans use their minds to train or learn—or even how dogs, dolphins, and other animals are trained to learn things.

So it is with neural networks. The lines the algorithm fit to the data are just lines fit to data using preset rules and it seems specious to call that training and learning. The algorithms do not know what they are manipulating, do not understand their results, and have no way of knowing whether the lines are meaningful or coincidental.

Oh, as you might have suspected, the two variables in Figures 1 and 2 were once again variables that I created using my computer's random number generator. Each value of each variable was determined by a

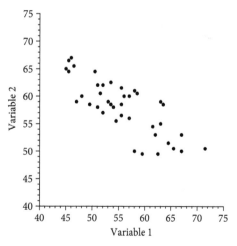

Figure 1 Training data?

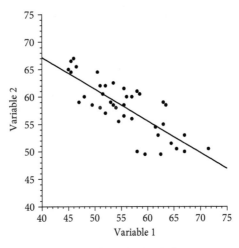

Figure 2 Machine learning?

random walk, independently of the other variable. The apparent statistical correlation is just a coincidence and means nothing at all. Of course, there is no way for a machine-learning algorithm to know this. Numbers are just numbers.

Earlier examples in Chapter 3 of how state-of-the-art neural networks misidentified seemingly random dots and patterns as a cheetah and starfish, and misidentified a young Middle Eastern male wearing colorful glass frames as Carson Daly were not intended to show that neural networks are hopeless, but rather to demonstrate that, despite their name, neural networks do not work the way human minds work.

As with other data-mining algorithms, there are inherent problems when neural network algorithms look for statistical patterns with no concern for logic or theory; for example, no concern for whether the two variables in Figure 1 and 2 *should* be related and no recognition of the fact that glasses do not change a person's face.

When used for complicated tasks, one huge drawback of deep neural networks is their opacity. No one knows exactly how they work and whether the results should be trusted. They are also very fragile. Image recognition software may identify a stop sign perfectly, but then be completely fooled by partly obscuring the sign or even by changing a single pixel.

Blinded by math

Stepwise regression, ridge regression, principal components, factor analysis, and neural networks are just five tools in a large tool box of statistical procedures used by data miners. I chose these five because they are representative and (relatively) simple to understand.

Data-mining tools, in general, tend to be mathematically sophisticated, yet naive in that they often make implausible assumptions. The problem is that the assumptions are hidden in the math and the people who use the tools are often more impressed by the math than curious about the assumptions.

During the Great Depression, when millions of people lost their jobs, their homes, their farms, and their businesses, the great British economist, John Maynard Keynes, criticized the prevailing classical economic models for assuming that people who lose their jobs choose to be unemployed:

The classical theorists resemble Euclidean geometers in a non-Euclidean world who, discovering that in experience straight lines apparently parallel often meet, rebuke the lines for not keeping straight as the only remedy for the unfortunate collisions which are occurring.

Even today, some economists respond to the criticism that their models are unrealistic by suggesting that the problem is with the world, not their models. Keynes dismissed such wishful thinking:

Professional economists, after Malthus, were apparently unmoved by the lack of correspondence between the results of their theory and the facts of observation . . . It may well be that the classical theory represents the way in which we should like our economy to behave. But to assume that it actually does so is to assume our difficulties away.

Classical economists were too enamored by the beauty of their models, and too little concerned with their realism.

So it is today with many of the statistical tools used by data miners.

CHAPTER 9

Take two aspirin

IBM's Watson got an avalanche of publicity when it won Jeopardy, but Watson is potentially far more valuable as a massive digital database for doctors, lawyers, and other professionals who can benefit from fast, accurate access to information.

A doctor who suspects that a patient may have a certain disease can ask Watson to list the recognized symptoms. A doctor who notices several abnormalities in a patient, but isn't confident about which diseases are associated with these symptoms, can ask Watson to list the possible diseases. A doctor who is convinced that a patient has a certain illness can ask Watson to list the recommended treatments. In each case, Watson can make multiple suggestions, with associated probabilities and hyperlinks to the medical records and journal articles that it relied on for its recommendations.

Watson and other computerized medical data bases are valuable resources that take advantage of the power of computers to acquire, store, and retrieve information. There are caveats though. One is simply that a medical data base is not nearly as reliable as a Jeopardy data base. Artificial intelligence algorithms are very good at finding patterns in data, but they are very bad at assessing the reliability of the data and the plausibility of a statistical analysis.

It could end tragically if a doctor entered a patient's symptoms into a black-box data-mining program and was told what treatments to use, without any explanation for the diagnosis or prescription. Think for a moment about your reaction if your doctor said,

I don't know why you are ill, but my computer says, "Take these pills."

or

I don't know why you are ill, but my computer recommends surgery.

Any medical software that uses neural networks or data reduction programs, such as principal components and factor analysis, will be hard-pressed to provide an explanation for the diagnosis and prescribed treatment. Patients won't know. Doctors won't know. Even the software engineers who created the black-box system won't know. Nobody knows.

Watson and similar programs are great as a reference tool, but they are not a substitute for doctors because: (a) the medical literature is often wrong; and (b) these errors are compounded by the use of data-mining software.

Call me in the morning

I had a routine physical checkup several years ago. My height and weight were measured; I was asked two pages of questions about my lifestyle (no, I don't smoke); and I was given a battery of tests. The nurse measured my temperature, heart rate, and blood pressure. Urine and blood samples were taken and tested for who knows what. That evening, I got a telephone call telling me that one test (I don't remember which one) came back worrisome. Ninety-five percent of all healthy patients have readings in a range that is labeled *normal*. One of my test results was outside the normal range, so I was evidently not healthy.

My doctor said, "Not to worry." She told me to take two aspirin, get a good night's sleep, and come back the next day to be retested. I did what I was told and, to my relief, my reading the next day was back inside the normal range.

Was it the aspirin or the good night's sleep? Probably neither. Most likely, it was just random noise. For virtually every test I took, there are variations in test results for perfectly healthy individuals. Blood pressure depends on the time of day, digestion, and a person's emotional state. Cholesterol can be affected by what a person has eaten and whether the person exercised before the test. Every test is prone to equipment errors and to human errors in reading, recording, and interpreting the results.

If a test result is abnormally high or low because of chance fluctuations, a second test will probably yield a result closer to the mean. This reversal makes it difficult to assess the value of medical treatments—in my case, whether the aspirin and good night's sleep had any effect at all.

It has been said that, "If properly treated, a cold will be gone in fourteen days; left untreated, it will last two weeks." This is part of the fundamental wisdom of the age-old advice offered by doctors who sound like they are trying to avoid being bothered when they say, "Call me in the morning."

Even if I *had* been ill and the aspirin had done nothing at all, I still might have felt better in the morning because of the body's remarkable ability to heal itself. Think of a scrape severe enough for you to bleed. The body clots the blood, forms a scab, and repairs the skin—all by itself, with no medical intervention.

These are two distinct reasons why "call me in the morning" works. First, medical tests are imperfect measures of a patient's condition. Second, people who are ill often improve as their bodies fight off what ails them.

The consequences are more serious than being worried unnecessarily. Abnormal readings due to chance fluctuations can lead to unnecessary treatments. The subsequent improvement in test results can lead to an unfounded belief that the treatment was effective.

Suppose that physical exams are given to a large number of people and those with the highest cholesterol readings are identified and given a special diet. We can expect their cholesterol readings to improve even if the dietary instructions are nothing more than, "Wave a hand over your food before eating it."

In addition, we all know some pain-relievers work better for some people and other pills work better for others. That's true of most medical treatments. They are neither 100 percent effective or 100 percent ineffective. When the effects are modest and vary from patient to patient, the results of a medical test depend on which persons are randomly assigned to the treatment group that is given the medication and which are put in the control group that is given a placebo.

Statisticians try to account for this random variation by estimating the probability that the differences between the treatment and control groups would be as large as observed if the differences were due solely to chance.

A standard level of statistical significance is five percent. This means that if the treatment being tested is worthless, there is only a five percent chance it will show statistically significant benefits. Fair enough, but this also means that five percent of all worthless treatments tested will show statistically significant results.

In the cutthroat world of medical research, brilliant and competitive scientists perpetually seek fame and funding to sustain their careers. To get fame and funding, they need to get published. To get published, they need to obtain statistically significant results—by any means necessary, including Texas Sharpshooter Fallacies #1 and #2.

Researchers can generate statistically significant results simply by testing lots of treatments. Even if they are so misguided as to test nothing but worthless remedies, they can expect five out of every hundred worthless treatments tested to turn out to be statistically significant—which is enough to generate published papers and approved grant proposals.

Similarly, pharmaceutical companies can make enormous profits from treatments that are clinically "proven" to be effective. One way to ensure that some treatments will be endorsed is to test thousands of treatments. No matter how many statistical hurdles are required, chance alone ensures that some worthless treatments will clear them all.

Let's look at three Texas sharpshooter examples.

I'll have another cup

In the early 1980s, *The New England Journal of Medicine*, one of the world's premier medical journals, reported that a group led by Brian MacMahon, a widely respected researcher and chair of the Harvard School of Public Health, found "a strong association between coffee consumption and pancreatic cancer." The Harvard group advised people to stop drinking coffee in order to reduce the risk of pancreatic cancer. MacMahon followed his own advice. Before the study, he drank three cups a day. After the study, he stopped drinking coffee.

One problem was Texas Sharpshooter Fallacy #1. The study was intended to investigate the link between drinking alcohol or smoking tobacco and developing pancreatic cancer. MacMahon looked at alcohol. He looked at cigarettes. He looked at cigars. He looked at pipes. When he didn't find anything, he kept looking. He tried tea. He tried coffee and finally found something: patients with pancreatic cancer drank more coffee.

If six independent tests are conducted, in each case involving something that is unrelated to pancreatic cancer, there is a 26 percent chance that at least one of these tests will find an association that is statistically significant at the 5 percent level—a 26 percent chance of making something out of nothing.

MacMahon's study had another flaw. He compared hospitalized patients who had pancreatic cancer to patients who had been hospitalized by the same doctors for other diseases. The problem was that these doctors were often gastrointestinal specialists, and many of their patients had given up coffee because of fears that it would exacerbate their ulcers. The patients with pancreatic cancer did not stop drinking coffee; so there were more coffee drinkers among the patients with pancreatic cancer. It wasn't that drinking coffee caused pancreatic cancer, but rather that patients who did not have pancreatic cancer stopped drinking coffee.

Subsequent studies, including one by MacMahon's group, failed to confirm the initial study. This time, they concluded that, "in contrast to the earlier study, no trend in risk was observed for men or women." The American Cancer Society agreed: "the most recent scientific studies have found no relationship at all between coffee and the risk of pancreatic, breast, or any other type of cancer."

More recent research has not only repudiated MacMahon's initial warning, it now appears—at least for men—that drinking coffee reduces the risk of pancreatic cancer!

Distant healing

In the 1990s a young doctor named Elisabeth Targ investigated whether patients with advanced AIDS could be healed by distant prayer and other positive thoughts. Forty AIDS patients were separated into two groups. Photos of those in the prayer group were sent to experienced distance healers (including Buddhists, Christians, Jews, and shamans) who lived an average of 1,500 miles from the patients. The other 20 patients were on their own.

The test was double-blind in that neither the Targ nor the patients knew who was being sent healing thoughts and who was being neglected—lest this knowledge might taint the results.

This six-month study found that those in the prayer group spent fewer days in the hospital and suffered fewer AIDS-related illnesses. The results were statistically significant and published in a prestigious medical journal. People with agendas cited Targ's work as proof of the existence of God or the inadequacy of conventional views of mind, body, time, and space.

Targ was awarded a $1.5 million grant from the National Institute of Health (NIH) for an even larger study of AIDS patients and for investigating whether distant healers could shrink malignant tumors in patients with brain cancer. Shortly after being awarded this grant, Targ was diagnosed with brain cancer herself and, despite being sent prayers and healing energy from all over the world, she died four months later.

After her death, problems were discovered with her 40-patient study. Targ had planned to compare the mortality of the prayer and non-prayer groups. However, one month after the six-month study began, triple-cocktail therapy became commonplace and only one of the 40 patients died—which demonstrated the effectiveness of the triple cocktail but eliminated the possibility of a statistical comparison of the prayer and non-prayer groups.

Targ and her colleague, Fred Sicher, looked for other differences between the two groups. They considered a variety of physical symptoms, quality-of-life measures, mood scores, and CP4+ counts. There were no differences between the prayer and non-prayer groups. Targ's father had done experiments attempting to prove that people have paranormal abilities to perceive unseen objects, read each other's minds, and move objects just by willing them to move. He told his daughter to keep looking. If you believe something, evidence to the contrary is not important. Just keep poking through the data for evidence that supports your beliefs. Finally, she found something—hospital stays and doctor visits, although medical insurance was surely a confounding influence.

Then Targ and Sicher learned about a paper listing 23 AIDS-related illnesses. Maybe they could find a difference between the prayer and non-prayer groups for some of these 23 illnesses. Unfortunately, data on these illnesses had not been recorded while the study was double-blind. Undeterred, Targ and Sicher pored over the medical records of their subjects even though they now knew which patients were in the control group and which were in the prayer group. When they were done, they reported that the prayer group fared better than the non-prayer group for some illnesses. This energetic data-mining seems to have been done without the benefit of data-mining software.

Their published paper suggested that the study had been designed to investigate the few illnesses that were found to be statistically significant (Texas Sharpshooter Fallacy #1) and did not reveal all the other tests that had been done or that the final data were assembled after the study was over and the double-blind controls had been lifted. Perhaps they found what they were looking for simply because they kept looking. Perhaps they found what they were looking for because the data were no longer double-blind.

Targ's NIH study continued after her death. It found no meaningful differences in mortality, illnesses, or symptoms between the prayer and non-prayer groups. An even larger study, conducted by researchers at the Harvard Medical School, looked at 1,800 patients who were recovering from coronary artery bypass graft surgery. They found no differences between those patients who received distant prayer and those that did not.

Cancer clusters

In the 1970s, Nancy Wertheimer, an epidemiologist, and Ed Leeper, a physicist, drove through Denver, Colorado, looking at homes that had been lived in by people who had died of cancer before the age of 19. They tried to find something—anything—these homes had in common. They found that many cancer-victim homes were near large power lines, so they concluded that exposure to the electromagnetic fields (EMFs) from power lines causes cancer.

A journalist named Paul Brodeur wrote three *New Yorker* articles that reported other anecdotal correlations between power lines and cancer. He ominously warned that, "Thousands of unsuspecting children and adults will be stricken with cancer, and many of them will die unnecessarily early deaths, as a result of their exposure to power-line magnetic fields."

The resulting national hysteria offered lucrative opportunities for consultants, researchers, lawyers, and gadgets, including Gauss meters that would allow people to measure EMFs in their own homes (rooms with high EMF readings were to be blocked off and used only for storage). Fortunately, the government did not tear down the nation's power lines.

The problem with this scare is that, even if cancer is randomly distributed among the population, data mining will, more likely than not, discover a geographic cluster of victims. To demonstrate this, I created a fictitious city with 10,000 residents living in homes evenly spaced throughout the city, each having a one-in-a-hundred chance of cancer. (I ignored the fact that people live in families and that cancer is related to age.) I used a computer random number generator to determine the cancer victims in this imaginary town. The resulting cancer map is shown

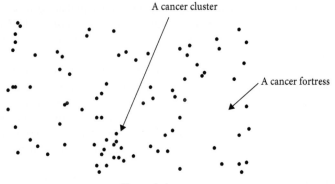

Figure 1 A cancer map

in Figure 1. Each black dot represents a home with a cancer victim. There are no cancer victims in the white spaces.

There is clearly a cancer cluster in the lower part of the map that would easily be discovered by any half-decent data-mining program. If this were a real city, we could drive through the neighborhood where these people live and surely find something special. Or data-mining software could ransack data looking for something unusual. Perhaps the city's Little League fields are nearby. If we now compare cancer rates for people who live near Little League fields with cancer rates for people who live far away, guess what? The cancer rates are higher near the fields, suggesting that living near a Little League field causes cancer.

Figure 1 also shows a cancer fortress, a part of town where nobody has cancer. There is surely something unusual in this cancer-free neighborhood that could be discovered by data-mining software or a Sunday drive. Perhaps the town's water tower is nearby. If we now compare the cancer rates for people who live near the water tower with cancer rates for people who live far away, cancer rates are lower near the water tower. That's precisely why we chose this neighborhood. Because nobody there has cancer.

In each case, near the Little League fields and near the water tower, we have the same problem—Texas Sharpshooter Fallacy #2. If we use the data to invent the theory (Little League fields cause cancer, water towers protect against cancer), then of course the data support the theory! How could it be otherwise? Would we make up a theory that did not fit the data?

An invented theory cannot be tested fairly by looking at the data that were used to invent the theory. We need fresh data. Other studies in other countries found no relationship between EMFs and cancer. Experimental studies of rodents found that EMFs far larger than those generated by power lines had no effect on mortality, cancer incidence, immune systems, fertility, or birth abnormalities.

Is there any theoretical basis for the power-line scare? Scientists know a lot about EMFs and there is no plausible theory for how power line EMFs could cause cancer. The electromagnetic energy is far weaker than that from moonlight and the magnetic field is weaker than the earth's magnetic field.

Weighing the theoretical arguments and empirical evidence, the National Academy of Sciences concluded that power lines are not a public health danger and there is no need to fund further research, let alone tear down power lines. One of the nation's top medical journals weighed in, agreeing that we should stop wasting research resources on this question.

In 1999, the *New Yorker* published an article titled "The Cancer-Cluster Myth," which implicitly repudiated the earlier articles written by Paul Brodeur. Nonetheless, the idea that cancer clusters are meaningful lives on. The internet has government-sponsored interactive maps that show the incidence of various cancers by geographic area, all the way down to census blocks. Millions of dollars are spent each year to maintain these maps with statistics that are up to date, but potentially misleading. One interactive site has cancer mortality rates for twenty-two types of cancer, two sexes, four age groups, five races, and more than three thousand counties. With millions of possible correlations, some correlations are bound to be frightening and easily discoverable by data-mining software.

To accommodate this fear, the U.S. Center for Disease Control and Prevention (CDC) has a web page where people can report cancer clusters they discover. Even though the CDC cautions that, "Follow-up investigations can be done, but can take years to complete and the results are generally inconclusive (i.e., usually, no cause is found)," more than a thousand cancer clusters are reported and investigated each year.

Most proven treatments don't work

Much (most?) published medical research is tainted by the two Texas sharpshooter fallacies: random variation in the data that is significant only if we overlook the fact that these flukes were uncovered by testing lots of theories, or inventing theories to match coincidental patterns in the data. The reported results vanish subsequently. This pattern is so common in medical research that it even has a name—the "decline effect."

Some researchers who have seen the decline effect first-hand with their own research are so perplexed that they set off on wild-goose chases looking for an explanation, when the reason is right in front of them. If the initial positive findings were due to a Texas sharpshooter fallacy, it is no surprise that the subsequent results are usually disappointing. It's just like our out-of-sample predictions of presidential elections based on temperatures in obscure cities. Statistical coincidences are fleeting.

A worthless treatment that appears to be effective is a false positive. There are also false negatives, cases in which an effective treatment does not show statistical significance. Consider a test with a five percent chance of a false positive—a five percent chance that the test will find a statistically significant difference between the treatment group and the control group when a worthless treatment is tested rigorously. Let's assume a ten percent chance of a false negative—a ten percent chance that an effective treatment will fail to show a statistically significant effect in a well-run test.

If there is only a five percent chance of a false positive and a ten percent chance of a false negative, it seems we should be able to tell the difference between worthwhile and worthless treatments almost every time. Not necessarily. It depends on how many of the tested treatments are effective and how many are worthless. Table 9.1 shows the implications if 1 percent of all tested treatments are effective and 99 percent are useless.

Of 10,000 treatments tested, 100 are effective; 90 of these 100 worthwhile treatments show statistically significant effects. Of the 9,900 worthless treatments tested, 495 generate statistically significant false positives. Overall, there are 585 tests with statistically significant results, but only 90 are really effective. An astonishing 85 percent of all treatments "proven" to be effective are actually worthless.

This paradox reflects a common confusion regarding inverse probabilities. One hundred percent of all players in the Premier League are male, but only a tiny fraction of all males play in the Premier League. Here, 90 percent of all effective treatments are statistically significant, but only 15 percent of all statistically significant treatments are effective.

These kinds of calculations are the basis for a famous paper with the provocative title, "Why Most Published Research Findings Are False,"

Table 9.1 *85 percent of all proven treatments are ineffective*

	Significant	Not significant	Total
Effective treatments	90	10	100
Ineffective treatments	495	9,405	9,900
Total	585	9,915	10,000

written by John Ioannidis, who holds positions at the University of Ioannina in Greece, the Tufts University School of Medicine in Massachusetts, and the Stanford University School of Medicine in California.

Ioannidis has devoted his career to warning doctors and the public about naively accepting medical test results that have not been replicated convincingly. His famous paper with the scandalous title works out the math as we did, although his assumptions are more damning than ours and the dismal probabilities even bleaker.

In addition to these theoretical calculations, Ioannidis has compiled a list of real-world "proven" treatments that turned out to be ineffective. In one study, he looked at 45 of the most widely respected medical studies published during the years 1990 through 2003. In only 34 cases were attempts made to replicate the original results with larger samples. The initial results were confirmed in 20 of these 34 cases (59 percent). For seven other treatments, the benefits were much smaller than initially estimated; for the final seven treatments, there were no benefits at all. Overall, only 20 of the 45 studies have been confirmed, and these were for the most highly respected studies! The odds are surely worse for the thousands of studies published in lesser journals. Ioannidis guesstimates that 90 percent of published medical research is flawed in that treatments reported to be beneficial are less beneficial than reported, sometimes worthless or worse.

Data mining

Traditional statistical tests assume that researchers start with well-defined theories and gather appropriate data to test their theories. Data mining works the other way around. Data come before theory, so test every theory you can test—whether the theories make sense or not.

If a medical treatment doesn't show statistical significance for the entire sample, see if it works for subsets. Separate the data by gender, race, age. Try different age categories. If the treatment doesn't work for the ailment you were studying initially, see if there are other beneficial effects.

Testing hundreds of treatments is an example of Texas Sharpshooter Fallacy #1: aim at hundreds of targets and only report the ones that were hit. Other medical research involves Texas Sharpshooter Fallacy #2: find a pattern and invent an explanation. It can happen in either the diagnosis or treatment of diseases.

First, the diagnosis. Suppose we have 100 patients who are known to have a certain disease and 100 who do not, and we record 1,000 characteristics for each person, perhaps blood tests, genetic information, race, hair color, eye color, residence. If we now use data-mining software to plunder this data base, we are sure to find some characteristics that are more common among the ill than among the well and are evidently good predictors of the disease.

For example, I was able to acquire a data base of systolic blood-pressure readings for 87 females. I also have complete information for 40 different characteristics of each patient. Some are numerical, like age, and some are categorical, like whether the person has ever smoked cigarettes.

I used data-mining software to see how well blood pressure can be predicted from these 40 characteristics. If my model works well, it might be used to identify other women who are at risk for high blood pressure. We can also identify high-risk factors (perhaps cigarette smoking) and recommend behavioral changes to reduce blood pressure for women with elevated values.

The model is quite successful, with a very respectable 0.72 correlation between actual and predicted blood pressure. Of the 23 women predicted to have systolic blood pressure above 130, 17 do. Figure 2 shows the predicted and actual values for all 87 women.

We can also get a respectable 0.47 correlation between predicted and actual blood pressure using just five of our patients' characteristics: 1, 12, 18, 23, and 34. So, doctors might focus on these five important characteristics for predicting and perhaps controlling elevated blood pressure.

What are these five important characteristics? Random numbers. I made up 87 phony women, using the number 87 so it would seem to be a real study. For 20 of the characteristics I used computerized coin flips to give values of 1 or 0, the same way that someone who smokes

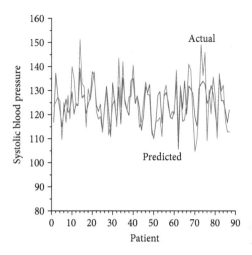

Figure 2 Predicted and actual systolic blood pressure

cigarettes might be coded 1 and someone doesn't is coded 0. For the other 20 characteristics, I had my computer generate normally distributed random variables with a mean of 100 and a standard deviation of 10. The phony blood-pressure readings were generated from a normal distribution with a mean of 125 and a standard deviation of 10. Every made-up characteristic of every made-up woman was created independently of the made-up characteristics of other women and independently of that woman's phony blood pressure and her 39 other made-up characteristics.

I filled this fictional medical data base with random numbers in order to make my point. Even if the characteristics recorded in a data base have nothing whatsoever to do with the presence or absence of the disease being analyzed, data-mining software may well discover statistically significant relationships that give the illusion of having discovered something useful.

The same is true of treatments. Suppose that a speculative treatment is administered to patients with a wide variety of medical conditions and data-mining software is used to identify those ailments or combinations of ailments that improved. Even if fluctuations in the patients' conditions are completely random, with no relationship whatsoever to whether or not they received the treatment, there are likely to be statistically significant patterns suggesting that the treatment works for some conditions. The distant-healing study is a good example of this fallacy.

Too much chaff, not enough wheat

Many miraculous treatments, like insulin and smallpox vaccines, have been discovered and proven effective through medical research. However, many published studies are flawed—usually because the data were ransacked to find something publishable.

This is an insurmountable problem for Watson and other medical advisory software. They are very good at collecting, storing, and retrieving medical data and journal articles—certainly better than any human. However, they have no common sense or wisdom. They do not know what the numbers and words mean and they have no way of assessing the relevance and validity of what is in their data bases. They do not know good data from bad data and they do not recognize when data have been tortured by the two Texas sharpshooter fallacies. They cannot distinguish between causation and coincidence, and they may even contribute to the problem by doing their own data-mining knowledge discovery.

All healthcare professionals are taught the precept, "First, do no harm." Experienced doctors maintain a healthy skepticism about medical research—taking a wait-and-see attitude about abstaining from coffee, relying on distant prayer, and tearing down power lines. They know about the pressure to publish and the decline effect. They are skeptical of black-box data-mining. My personal doctor scoffed at the idea of trusting a black-box algorithm to prescribe a medication or medical treatment.

Medical software programs can help doctors, but they cannot replace them.

CHAPTER 10

Beat the market I

Back when scam artists sent snail-mail instead of e-mail, I received a letter that began "Dear friend," a clear sign it was from someone trying to sell me something. Nonetheless, I read a bit more and saw this line highlighted in yellow: "IMAGINE turning $1,000 into $34,500 in less than one year!" Real friends don't highlight their sentences, but I pushed on, thinking I might share this BS with my students. Sure enough, it was a con.

The letter said that "no special background or education" was needed and that, "It's an investment you can make with spare cash that you might ordinarily spend on lottery tickets or the race track." Now I wasn't sure that I wanted to share this letter, lest my students wonder where this company had gotten my name. I don't buy lottery tickets or bet on horses. What had I done that made this company think I was a sucker?

The letter claimed that, instead of wasting my money on lottery tickets and horse races, I could get rich buying low-priced stocks. For example, the price of LKA International had jumped from 2 cents a share to 69 cents a share in a few months, which would have turned $1,000 into $34,500. All I had to do was pay $39 for a special report that would give me access to "the carefully guarded territory of a few shrewd 'inner circle' investors."

The entire premise is ridiculous. If someone really knew how to turn $1,000 into $34,500, they would be doing it, instead of selling special reports for $39. Yet, we repeatedly fall for such scams because we are hard-wired to think that the world is governed by regular laws that we can discover and exploit. Stock prices cannot be random. There must be underlying patterns, like night and day, winter and summer. Our gullibility is aided and abetted by our greed—by the notion that it is easier to make money by buying and selling stocks than by being a butcher, baker, or candlestick maker.

The inconvenient truth is that zigs and zags in stock prices are mostly random, yet transitory patterns can be found in random numbers. If we look for them, we will find them and be fooled by them.

Noise

People are often lured to the stock market by the ill-conceived notion that riches are there for the taking—that they can buy a stock right before the price leaves the launchpad and sell the stock before it crashes. They see stocks like LKA, think about how much money they could have made, and dream about how much money they will make.

Looking backward, it is easy to find missed opportunities. Looking forward, it is hard to predict which stocks will go up and which will go down. The stock market doesn't give away money.

The stock market is not a computer program that sets prices by following rules. Stock prices are determined by investors making voluntary transactions—some buying and others selling. Investors won't sell a stock for 2 cents a share if it is clear that the price will soon be 69 cents. Even if only a shrewd inner circle know that the price will soon be 69 cents, they will buy millions of shares, driving the price today up to 69 cents.

When LKA traded at 2 cents, there were as many buyers as sellers, as many people who thought that the stock was overpriced as thought it was a bargain. It is never evident that a stock's price is about to surge or collapse—for if it was, there wouldn't be a balance between buyers and sellers.

There are two reasons why it is hard to predict stock prices. One is that a stock's price changes when there is new information—when investors learn something they did not already know. Since new information, by definition, cannot be predicted, changes in stock prices caused by new information cannot be predicted. A second reason is that buyers and sellers are affected by human emotions and misperceptions that change in unpredictable, sometimes irrational, ways—what Keynes called "animal spirits."

Millions of investors have spent billions of hours trying to discover a formula for beating the stock market. It is not surprising that some have stumbled on rules that explain the past remarkably well but do poorly predicting the future. Many such systems would be laughable, except for the fact that people believe in them. What these wacky theories have in common is they are based on data mining.

The most dangerous trading systems are black-box models, where sophisticated statistical analyses, like principal components and factor analysis, are used to unearth patterns that are too complicated to understand, and it is impossible to determine whether the patterns make sense. We must take it on faith that the data mining has uncovered something real—not fool's gold.

By definition, we don't know what is inside black-box investment models, but we can learn some lessons by looking at data-mined systems that are not hiding inside black boxes.

Wacky theories

Long heels on women's shoes are supposed to be bullish and short heels bearish. Thin ties on males are thought to be bullish and wide ties bearish, though the *Wall Street Journal* noted that "some analysts put it the other way around while those who shop at Brooks Brothers aren't aware that tie widths can change."

Analysts have monitored sunspots, the water level of the Great Lakes, and sales of aspirin and yellow paint. Some believe that the market does especially well in years ending in five—1975, 1985, and so on—while others argue that years ending in eight are best. Burton Crane, a long-time *New York Times* financial columnist, reported that a man "ran a fairly successful investment advisory service based on his 'readings' of comic strips in *The New York Sun*." *Money* magazine once reported that a Minneapolis stock broker selected stocks by spreading the *Wall Street Journal* on the floor and buying the stock touched by the first nail on the right paw of his golden retriever. The fact that he thought this would attract investors says something about him—and his customers.

In 1987, a year with three Friday the 13ths, the chief economist at a Philadelphia bank reported that in the past 40 years there had been six other years with three Friday the 13ths, and a recession started in three of those years. Which is worse, the idea that an economist wasted his time on such nonsense or the fact that he was paid for wasting his time? Somehow, 1987 escaped without a recession.

Perhaps there is a competition for silly advice. Back before it went bankrupt, the L.F. Rothschild investment bank reported that during the past six Dragon years in the Chinese calendar, the U.S. stock market had gone up four times and down twice.

One of my favorites is the Boston Snow (B.S.) indicator, which uses the presence or absence of snow in Boston on Christmas Eve to predict the stock market the following year. An analyst at Drexel Burnham Lambert reported that "the average gain following snow on the ground was about 80% greater than for years in which there was no snow."

The Super Bowl Stock Market Predictor says that the stock market goes up if the National Football Conference (NFC) or a former National Football League (NFL) team now in the American Football Conference (AFC) wins the Super Bowl; the market goes down otherwise. A Green Bay Packer win is good for the stock market; a New York Jets win is bad for stocks. The man who created the Super Bowl Indicator intended it to be a humorous way of demonstrating that correlation does not imply causation. He was flabbergasted when people started taking it seriously! In 2017 I was told that some big fund managers still take it seriously. As I've said over and over, some people think that patterns are persuasive. Period.

In 2014 a former student forwarded me a trading rule concocted by an ex-chairman of Goldman Sachs Asset Management:

There's an odd little thing called the five-day rule. . . . The rule simply states that when the main U.S. stock-market indexes show a combined positive return after the first five days of trading, the year as a whole is very likely to be a good one. . . .

Jose Ursua, a former colleague of mine at Goldman Sachs, has run these numbers all the way back to 1928. He finds that when stocks rallied during the first five days, there was a 75.4 percent chance of a rally for the year. For the period since 1950, the probability rises to 82.9 percent. Few rules in finance are as unambiguous as that.

First, this sure sounds like tortured data. (Why five days? Why 1928?) Second, the first 5 trading days are part of the year's (approximately) 250 trading days, so that biases the calculations. He should have compared the first 5 days with the other 245 days. Third, the incorrect impression is that it is 50–50 whether the market goes up. In fact, the market usually goes up. It went up 73 percent of the years 1928–2013, 78 percent of the years

1950–2013. The five-day rule doesn't increase the odds much, especially if we consider that those five days are part of the year.

These investment rules are so old-school—worthless patterns uncovered by humans spending countless hours sifting through data. However, Friday the 13ths, Dragon years, and other nonsense are *exactly* the sort of nonsense that might be uncovered by a data-mining program looking for patterns—only computers are much more efficient at discovering worthless patterns.

Sometimes, an uncovered pattern has an aura of plausibility, but is clearly the useless fruit of tortured data. For example, it makes sense that the Federal Reserve's monetary policies affect the stock market. Low interest rates make it less expensive for consumers and businesses to borrow money. Low interest rates make stocks more attractive than bonds to investors. Believing that changes in the money supply have an effect on the stock market, Beryl Sprinkel, a prominent economist who was once chair of the Council of Economic Advisors, looked for a relationship and found one. Which index of stock prices? He chose the S&P 425 Industrial Index instead of the usual S&P 500. Which measure of the money supply? He chose a very narrow definition called M1.

From a visual inspection of graphs of the S&P 425 and M1, he concluded that, although the relationship is highly variable, "changes in monetary growth lead changes in stock prices by an average of about 15 months prior to a bear market and by about 2 months prior to bull market." Notice the convenient flexibility provided by *changes in monetary growth*. Over the past week? Month? Year? Sprinkel chose a 6-month horizon, no doubt after trying several other possibilities. Notice, also, the ill-defined terms *bull market* and *bear market*. Nobody knows when bull markets and bear markets begin and end.

With further twisting and tweaking, he discovered that the following strategy would have given an average profit of 6 percent a year between 1918 and 1960, while simply buying and holding stocks would have yielded 5.5 percent a year:

a Keep track of the average monthly rate of growth of seasonally adjusted M1 over the preceding 6 months.
b Buy stocks 2 months after a trough in this moving average.
c Sell stocks 15 months after a peak in this moving average.

All that effort and an extra 0.5 percent was all he could come up with? Nowadays, a computer data-mining algorithm would surely do better.

Sprinkel's convoluted model sure sounds like data torturing. Why start in 1918? Why the S&P 425? Why M1? Why the average monthly growth rate over a six-month period? Why 2-month and 15-month lags? I hope it doesn't surprise you that later studies found that the money supply is of no value in predicting changes in stock prices.

When Sprinkel discovered this worthless rule, he no doubt had to work really hard to make the necessary calculations and draw his graphs, fiddling with the specification until he found something that fit the data pretty well. Nowadays, unfortunately, computers can find such worthless rules very, very quickly.

Technical analysis

What's so hard about predicting the stock market? Anyone with open eyes can see patterns in stock prices—patterns that foretell whether prices are headed up or down.

Thirty years ago, a former student named Jeff called me with exciting news—my lectures on the folly of trying to discern profitable patterns in stock prices were hogwash. Jeff had taken a job with IBM and, in his spare time, he studied stock prices and had found some clear patterns. He was fine-tuning his system, and would soon be rich. He told me that he was going to rent a helicopter and land it on the lawn outside my classroom, so that he could march into my investments class triumphantly and tell students the truth.

Every year, I tell students this story. Then I walk over to a classroom window and look outside to see if Jeff's helicopter is parked on the lawn. I'm still waiting.

Technical analysts try to gauge investor sentiment by studying stock prices, trading volume, and various measures of investor moods. Technicians do not look at dividends or profits, either for individual companies or for the market as a whole. If they are studying an individual company, they do not need to know the company's name. It might bias their reading of the charts. This is precisely what data-mining algorithms do: look for patterns in numbers without any consideration of what the numbers represent.

John Magee, who co-authored the so-called bible of technical analysis, boarded up the windows of his office so that his readings of the hopes and fears of the market would not be influenced by the sight of birds singing or snow falling. He was inside his own black box.

A technician's most important tool is a chart of stock prices. The most popular are vertical-line charts, traditionally using daily price data. Each vertical line spans the high and low prices, with horizontal slashes showing the opening and closing prices. A technician adds lines, like the channel in Figure 1, to show a trend or other exploitable pattern. (To reduce the clutter, I omitted the vertical lines and just show the closing prices.)

An inverted head and shoulders

The core of technical analysis is identifying patterns in past stock prices that can be used to predict future prices. The analysis is made to sound legitimate by affixing labels to these patterns, such as a channel, support level, resistance level, double top, double bottom, head and shoulders, cup and handle. Despite the alluring labels, study after study has found that technical analysis is pretty much worthless—except for employing technical analysts and generating commissions for stockbrokers.

An academic economist once sent several stock-price charts, including Figures 1, 2, and 3 (without the lines), to a technical analyst (let's call

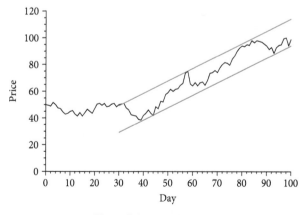

Figure 1 An upward channel

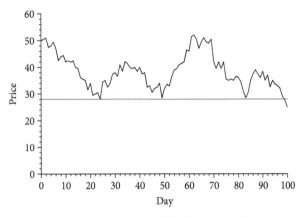

Figure 2 Piercing a head-and-shoulders support level

him Ed) and asked Ed's help in deciding whether any of these stocks looked like promising investments. The economist did not identify the companies, and Ed did not ask.

Ed drew the two parallel lines in Figure 1 and saw a clear pattern. Since about Day 30, this stock had traded in a narrow upward-sloping channel. On Day 100, the price was near the lower boundary of this channel and clearly poised for an upward surge.

Figure 2 shows a stock with a support level that was established on Day 24 and confirmed twice. Every time the price fell to $28, it found support and rebounded upward. Even more powerfully, this chart shows a head-and-shoulders pattern, with the price rising off the support level at $28, falling back to $28, rising even more strongly, falling back to $28, rising modestly, and retreating to $28 a third time.

Like many technical analysts, Ed believed that support levels that are established and confirmed by a head-and-shoulders pattern are extremely strong. When the price pierced the $28 support level on Day 99, this was an unmistakable signal that something had gone terribly wrong. It takes an avalanche of bad news to overcome such a strongly established support level. Once the $28 barrier had been breached, the price could only go down.

Figure 3 shows the opposite pattern. A resistance level was established and confirmed at $65. Every time the price approached $65, it

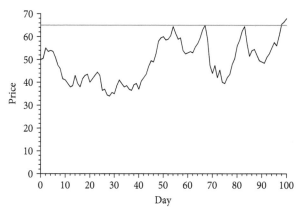

Figure 3 Breaking through a resistance level

bounced back. The more often this occurred, the stronger was the psychological resistance to the price going above $65. Even more powerfully, this chart shows an inverted head-and-shoulders pattern, as the second bounce back from $65 was much stronger than the first and third bouncebacks. When the price surged through the resistance level on Day 98, eliminating the psychological barrier to the price going higher, this was an unmistakable buy signal.

Ed was so excited by the patterns he found in these three charts that he overlooked the odd coincidence that all three stock prices started at $50 a share. This was not a coincidence.

These are not real stocks. The mischievous professor (yes, it was me) created these fictitious data from student coin flips in one of my investments classes. In each case, the "price" started at $50. Then, each day's price change was determined by 25 coin flips, with the price going up 50 cents if the coin landed heads and down 50 cents if the coin landed tails. For example, 14 heads and 11 tails would be a $1.50 increase that day. After generating dozens of charts, I sent ten to Ed with the expectation that he would find seductive patterns. Sure enough, he did. So would data-mining software.

When this ruse was revealed, Ed was quite disappointed that these were not real stocks, with real opportunities for profitable buying and selling.

However, the lesson he drew from this hoax was quite different from the intended lesson: Ed concluded that it is possible to use technical analysis to predict coin flips!

It is very difficult, even for seasoned professional investors, to understand that random noise can generate patterns that are nothing more than chance. When we see a pattern, it is tempting to think that it must be meaningful and hard to believe that it may be coincidental.

Flipping coins

I sometimes do another coin-flipping experiment in my investment classes to demonstrate this principle. I announce that I am going to flip a coin several times and see if anyone can predict the outcomes. Before my first flip, I use my hand to divide the class into the left and right sides of the room. The left side will predict heads, the right side tails. If there are 32 students in the class, 16 are bound to be correct. If my first flip lands tails, the 16 students on the right side of the room were correct. I now divide these 16 students into 8 in front and 8 in back. The front 8 predict heads for the second flip; the back 8 go with tails. If the second flip is tails, the 8 students in back have now predicted two flips correctly. I divide these 8 into two groups and make my third flip. The 4 students who were right are divided again and the fourth flip takes us down to 2 students who have called all four flips correctly. One more flip and we have our winner— the student who correctly predicted five coin flips in a row. I then ask if anyone in the class wants to bet that this student will predict the next five flips correctly. No one does.

This is data mining. There are 32 coin-flip predictors, just as there might be 32 systems for predicting stock prices. After the fact, I discovered the most successful coin-flip predictor, just like discovering the most successful stock-predicting system. In either case, identifying a successful predictor after the fact proves nothing at all because there will always be a most successful predictor, even if nothing more than luck is involved.

Going forward, the student who is the most successful coin-flip predictor is nothing special, nor is a lucky stock market system. The successful coin-flip predictor would be shown to be unskilled by being tested on another five flips, and so might a successful stock system be shown to be useless by being tested with fresh data.

However, if enough coin flippers and worthless stock market systems are tested, some will inevitably be successful through both rounds of testing. If I had a bigger class, 1,024 students to be exact, I could discover 32 students who correctly predict the first five flips. If these 32 students are tested on five more coin flips, one will get the second five flips correct, giving ten correct predictions in all. Despite having been tested and retested, this student still has no special ability when it comes to predicting going flips. Data mining 1,024 students who predict 10 coin flips is no more conclusive than data mining 32 students who predict five coin flips.

So it is with stock market systems. Data mining software might examine millions, billions, or trillions of potential stock-predicting systems and identity some that would have been very successful. If these systems are retested, some will continue to be successful—even if there is nothing more to them than the kind of luck demonstrated by students predicting coin flips. No matter how many times useless stock market systems are tested and retested, some will pass if enough are tested.

It is instructive to look at several trading rules that have been used and recommended by serious people so that we can see how easy it is to be fooled by systems that explain the past remarkably well and predict the future remarkably poorly.

Wall Street Week's ten technical indicators

For more than 30 years, *Wall Street Week* was one of the Public Broadcasting Systems most popular television shows. Hosted by Louis Rukeyser, the show aired on Friday evenings after the New York Stock Exchange closed. The show began with Rukeyser's recap of the week on Wall Street, enlivened by corny jokes and groan-inducing puns. When a viewer sent in a letter asking about investing in a toupée maker, Rukeyser answered, "If all your money seems to be hair today and gone tomorrow, we'll try to make it grow by giving you the bald facts on how to get your investments toupée."

For many years, one segment of the show reported the latest readings of ten technical indicators developed by Robert Nurock, a broker, investment consultant, and regular panelist on the show. I subscribed to his newsletter for several years because I wanted to see how his indicators evolved over time as the patterns he discovered appeared and then evaporated.

His ten selected indicators first appeared in 1972 and were revised at least five times over the next 17 years. Nurock's interpretation of his indicators was generally contrarian: when the stock market seemed strong, he advised selling; when the market was weak, he advised buying. That is usually pretty good advice, but the idea that there are reliable predictors of *when* stock prices will go up or down is ill-advised.

Nurock's indicators are interesting because they are representative of the many barometers that technicians watch—and their evolution is instructive. I will look at three.

Market breadth

Many analysts use an advance-decline index to monitor the number of stocks advancing relative to the number declining. Unlike the Dow Jones Industrials, S&P 500, and other market indexes, such a comparison pays no attention to the value of the stock or to the size of the advance or decline—a 25-cent change in Fly-By-Night Marketing counts the same as a five-dollar change in IBM. That is why it is called a measure of *market breadth*.

Wall Street Week initially used an advance-decline index that was a five-week average of the daily ratio of the number of stocks advancing plus 50 percent of those unchanged to the total number traded. A value of 0.50 shows that the market is mixed, with the number of stocks increasing equal to the number declining. Consistent with a contrarian approach, Nurock interpreted a large value for this index (when most stocks are rising) as evidence that you should sell, and a low value (when most stocks are falling) as a signal to buy.

Nurock's specific cutoff values depended on whether a bull or bear market prevailed, as shown in Table 10.1. The bull/bear distinction is discomforting,since we are given no objective way to gauge which is the

Table 10.1 *Nurock's advance–decline index, 1972*

	Buy signal	Neutral	Sell signal
In a bull market:	43 to 45 percent	45 to 57 percent	57 to 60 percent
In a bear market:	38 to 40 percent	40 to 50 percent	50 to 54 percent

case at any point in time. A few years later, the 38-to-40 bear-market buy signal was revised to 36-to-40 percent; later, the index was dropped entirely.

The ambiguities in interpreting technical indicators are also demonstrated vividly by the fact that Louis Rukeyser gave an explanation of Nurock's advance–decline index that is the opposite of Nurock's interpretation. Here is Rukeyser's description (including a trademark pun):

Practically every technician uses some variant of this, which tots up how many stocks went up versus how many stocks went down. The result is a measure of the market's "breadth"—and bad breadth stinks. If the averages are rising but breadth is declining, technicians will assume that the market is already beginning to deteriorate.

Some technicians do draw this conclusion but, contrary to Rukeyser, Nurock did not. When most stocks were falling, Nurock gave a buy signal.

The first revision of Nurock's ten indicators added a second advance–decline index: a ten-day average of the number of stocks advancing minus the number declining. A value below −1000 (showing that most stocks have been falling) was bullish and a value above +1000 bearish. This indicator survived the second revision, but the reading was made considerably more complicated. The bullish signal became "an expansion of this index from below +1000 to the point where it peaks out and declines 1000 points from this peak."

As with the other technical indicators, Nurock clearly made up a model to fit past patterns in the data. When the data-mined model did not work with fresh data, he changed the model. When the new data-mined model did not work, he made up a new model. Make up model. Test. Change model. Repeat.

Low-price activity ratio

Another interesting *Wall Street Week* indicator, similar to those followed by many technicians, is the low-price activity ratio, which divides the weekly ratio of the volume of trading in Barron's Lo-Price Stock Index by the volume of trading in the Dow Jones Industrials. The idea is that low-priced stocks are more speculative than the blue chips included in the Dow Jones average and, so, a relative increase in their trading indicates more speculative activity.

In the reaction spirit, a low value of this indicator is a signal to buy and a high value is a signal to sell. The particular values used initially by *Wall Street Week* were:

Buy signal	Neutral	Sell signal
5 to 8 percent	8 to 10 percent	10 to 15 percent

In the first revision, these values were changed to:

Buy signal	Neutral	Sell signal
5 to 7 percent	7 to 12 percent	over 12 percent

In the second revision, these were changed once again, to:

Buy signal	Neutral	Sell signal
below 12 percent	12 to 18 percent	over 18 percent

Then the third revision:

Buy signal	Neutral	Sell signal
below 4 percent	4 to 8 percent	over 8 percent

Make up model. Test. Change model. Repeat.

Advisory service sentiment

The first revision of *Wall Street Week*'s technical indicators introduced another interesting technical indicator: advisory service sentiment. There are a wide variety of investment services that sell their recommendations to investors. On average, these self-proclaimed experts tend to be bullish, though the fraction that are bullish varies with the market. One advisory service, *Investor's Intelligence*, subscribes to other leading services and provides its subscribers with a weekly tabulation of the fraction that are bearish. Interestingly, its interpretation (and Nurock's) is that the more bearish are the advisory services, the more likely it is that the stock market will go up.

Nurock explains: "When advisory service sentiment becomes overly one sided it is viewed as a contrary indicator as services tend to follow trends rather than anticipate changes in them." Another explanation is that investment advisors are amateurs who are wrong more often than right. If they're so smart, why aren't they using their own advice rather than selling it? Nurock's sentiment indicator initially signaled sell if the percent bearish was less than 15 percent and buy if the percent bearish was over 30 percent in a bull market or over 60 percent in a bear market. These were later changed substantially, to sell if the percent bearish was below 25 percent and buy if above 42 percent. Then the cutoffs were made misleadingly precise: 35.7 percent and 52.4 percent. That's the nature of data mining. Make up model. Test. Change model. Repeat.

Tweet, tweet

In 2011, a team of researchers reported that a data-mining analysis of nearly 10 million Twitter tweets during the period February to December 2008 found that an upswing in "calm" words predicted an increase in the Dow Jones average up to six days later. No, I am not making this up. This is the kind of provocative research that gets reported worldwide even though the data may be completely unreliable and the claims are utterly implausible.

People who tweet are not a random sample of the population, let alone investors. Many people do not have Twitter accounts. Some have accounts, but don't log in much. Some have accounts and log in, but seldom tweet. (Twitter estimates that nearly half the people who log in are listeners, not tweeters.) Some tweeters are not real people, but automated bots.

Even the lead author admitted that he had no explanation for the observed pattern, especially since many tweets are from teenagers and people living outside the United States. How could the Dow industrial average be affected by a teenager sharing what she had for breakfast or a German commenting on a Bayern Munich goal? It surely isn't, but they might be correlated statistically.

Let's start with the fact that these Texas sharpshooter researchers looked at seven different predictors: an assessment of positive versus negative moods and six mood states (calm, alert, sure, vital, kind, and happy). They also considered several different days into the future for correlating with the Dow. The reported research also suggests some flexibility in selecting the words that represent the seven different predictors. Finally, why did they use data from February to December 2008? What happened to January? Why did a 2011 paper use 2008 data? Did the discovered patterns only exist during that peculiar period?

I have one more question for the authors: Did they start trading stocks based on their reported findings?

Technical gurus

The Wall Street adage, "There is nothing wrong with the charts, only the chartists," confirms the difficulty in finding useful signals. Nonetheless, technical analysts are employed by all major brokerage firms and many operate their own advisory services. Periodically, a technical analyst is elevated to the status of financial guru when reports of astoundingly accurate predictions are recounted in the media and devoted followers avidly seek the advice of these celebrities.

The Elliot wave theory is one example. A thirteenth century Italian, Leonardo Fibonacci, investigated the progression 1, 1, 2, 3, 5, 8, 13, 21, 34, 55, 89, . . . , now called a Fibonacci series, where each number is equal to the sum of the two preceding numbers. The number of branches on a tree grows according to a Fibonacci series, as does the family tree of a male bee and the diameter of spirals on a seashell. Musical scales conform to a Fibonacci series, and the Egyptians used Fibonacci series in designing the Great Pyramid at Giza.

The Elliot Wave theory is Ralph Nelson Elliot's application of Fibonacci series to stock prices. Elliott was an accountant, with an accountant's fascination with numbers. After studying ups and downs in stock prices, he concluded that these ups and downs are the complex result of several overlapping waves:

> Grand waves that last for centuries
> Supercycles that last for decades
> Regular cycles that last for years
> Primary waves that last for months
> Intermediate waves that last for weeks or months
> Minor waves that last for weeks
> Minute waves that last for days
> Minuette waves that last for hours
> Subminuette waves that last for minutes.

Elliott proudly proclaimed that "because man is subject to rhythmical procedure, calculations having to do with his activities can be projected far into the future with a justification and certainty heretofore unattainable." To be certified as a Chartered Market Technician by the Market Technicians Association, one must pass a test that includes questions about the Elliott Wave theory.

Yes, the wave theory is complicated, but that is part of its appeal. Those who are mathematically inclined are attracted to its complex elegance, a complexity that gives it the flexibility to fit virtually any set of data, even random coin flips. It is a perfect data-mining system, more easily implemented today with computers.

After the fact, wave enthusiasts can always come up with a wave interpretation, though different enthusiasts often have different explanations. Before the fact, wave theorists often disagree and are often wrong, though they are adept at explaining why afterward.

In the 1980s, Robert Prechter became the most famous Elliott Wave disciple. In March 1986, *USA Today* called Prechter the "hottest guru on Wall Street," after a bullish forecast he made in September 1985 came true. The same article advised readers of his forecast that the Dow would rise to 3600–3700 by 1988; however, the high for 1988 turned out to be 2184. In October 1987, Prechter said, "The worst case [is] a drop to 2295," just days before the Dow collapsed to 1739. *The Wall Street Journal* published a front-page article in 1993 with the headline, "Robert Prechter sees his 3600 on the Dow – But 6 years late." The Dow hit 3600, just as he predicted, but six years after he said it would. Prechter should have heeded the advice given me by a celebrated investor about how difficult it is to predict stock prices: "If you give a number, don't give a date."

The Foolish Four

In 1996 two brothers, David and Tom Gardner, launched what has become a small empire with *The Motley Fool Investment Guide: How the Fools Beat Wall Street's Wise Men and How You Can Too*. They called one of their strategies The Foolish Four. They invented this strategy by studying data for the years 1973–1993. If they had known of the strategy beforehand (Ha! Ha!), they would have earned an average annual return of 25 percent.

They advised readers that their system "should grant its fans the same 25 percent annualized returns going forward that it has served up in the past."

Here's their Foolish Four Strategy:

1 At the beginning of the year, calculate the dividend yield for each of the 30 stocks in the Dow Jones Industrial Average. For example, on December 31, 2016, Coca-Cola stock had a price of $41.46 per share and paid an annual dividend of $1.48 per share. Coke's dividend yield was $1.12/$41.46 = 0.0357, or 3.57 percent.
2 Choose the ten stocks with the highest dividend yields.
3 Of these ten stocks, choose the five with the lowest price per share.
4 Of these five stocks, cross out the one with the lowest price.
5 Invest 40 percent of your wealth in the stock with the next lowest price.
6 Invest 20 percent of your wealth in each of the other three stocks.

I hope your mind is throbbing: "Data mining!"

The Foolish Four Strategy is so convoluted and so little related to anything sensible, we know for a fact that it was the product of extensive data mining.

Two finance professors thought the same thing, so they tested the Foolish Four Strategy for the years 1949–1972, just prior to the years data-mined by the Gardners. Looking backwards, the professors found it was a flop. Looking forward, the Gardners found it was a flop. In 1997, one year after launching the Foolish Four Strategy, the Gardners tweaked their system and renamed it the UV4: "Why the switch? History shows that the UV4 has actually done better than the old Foolish Four." Make up model. Test. Change model. Repeat.

The Gardners broke the cycle in 2000 when they stopped touting the Foolish Four and UV4 strategies. No surprise, if your initial reaction was, "Data-mining!"

Black box investing for fun and profit

The value of technical analysis seems self-evident. Any reasonably alert person can see well-defined patterns in stock prices. However, investors need a crystal ball and stock charts provide a rear view mirror. Ransacking data for patterns demonstrates little more than the researcher's persistence. Remember, "If you torture the data long enough, it will confess."

There are two lessons. First, we should recognize the temptation to look for patterns and think that the patterns we find must be meaningful—

despite all evidence to the contrary. Second, we should strive to resist our innate susceptibility.

Technical analysis used to be done by humans scrutinizing charts, searching for patterns. Now, computers can be programmed to look for patterns that are too complicated and subtle to be detected by visual inspection, but may be just as worthless—which is why it good to remember those two lessons. It is tempting to think that patterns must be meaningful. It is wise to resist that temptation.

CHAPTER 11

Beat the market II

Nowadays, technical analysts are called quants. Being overly impressed by computers, we are overly impressed by quants using computers instead of pencils and graph paper.

Quants do not think about whether the patterns they discover make sense. Their mantra is, "Just show me the data." Indeed, many quants have PhDs in physics or mathematics and only the most rudimentary knowledge of economics or finance. That does not deter them. If anything, their ignorance encourages them to search for patterns in the most unlikely places.

The logical conclusion of moving from technical analysts using pencils to quants using computers is to eliminate humans entirely. Just turn the technical analysis over to computers.

A 2011 article in the wonderful technology magazine *Wired* was filled with awe and admiration for computerized stock trading systems. These black-box systems are called *algorithmic traders* (*algos*) because the computers decide to buy and sell using computer algorithms in place of human judgment. Humans write the algorithms that guide the computers but, after that, the computers are on their own.

Some humans are dumbstruck. After Pepperdine University invested 10 percent of its portfolio in quant funds in 2016, the director of investments argued that, "Finding a company with good prospects makes sense, since we look for undervalued things in our daily lives, but quant strategies have nothing to do with our lives." He thinks that not having the wisdom and common sense acquired by being alive is an argument for computers. He is not alone. Black-box investment algorithms now account for nearly a third of all U.S. stock trades.

Some of these systems track stock prices; others look at economic and noneconomic data and dissect news stories. They all look for patterns. A momentum algorithm might notice that when a particular stock trades at a higher price for five straight days, the price is usually higher on the sixth day. A mean-reversion algorithm might notice that when a stock trades at a higher price for eight straight days, the price is usually lower on the ninth day. A pairs-trading algorithm might notice that two stock prices usually move up and down together, suggesting an opportunity when one price moves up and the other doesn't. Other algorithms use multiple regression models. In every case, the algorithm is based on data mining and the motto is, "If it works, trade on it."

I invest and I teach investments, so I decided to try a little data mining myself to see if I could find a reliable predictor of stock prices. With any luck, my data mining might yield a "knowledge discovery" and make some money.

The stock market and the weather

It has been reported that the weather in New York City affects the U.S. stock market, though the effects may have weakened over time as trading has evolved from trading floors to electronic orders from all over the country and the world.

Heidi Artigue had collected the daily high and low temperatures in 25 cities, so it was tempting to see if I could find some temperatures that would explain the daily fluctuations in the S&P 500 index of stock prices.

My initial thought was that daily temperatures are of limited use for forecasting stock prices because temperatures have seasonal patterns, but stock prices do not. In addition, stock prices have a pronounced upward trend, but temperatures do not (at least over short periods like a few years). Nonetheless, with very little data mining, I was able to find five temperatures that did a reasonable job of forecasting stock prices in 2015.

The high and low temperatures in 25 cities gave me 50 potential explanatory variables. These were the basis for 50 possible models with one explanatory variable, 1,225 models with two explanatory variables, 19,600 with three variables, 230,300 with four variables, and 2,118,760 with five variables. Acting as a dedicated data miner, I estimated them all, a total of 2,369,935 models.

Many were pretty good, but the best was:

$$Y = 2361.65 - 3.00C + 2.08M - 1.85A + 1.98L - 3.06R$$

where:

$C =$ Curtin, Australia, high temperature

$M =$ Omak, Washington, low temperature

$A =$ Antelope, Montana, high temperature

$L =$ Lincoln, Montana, low temperature

$R =$ Rock Springs, Wyoming, low temperature

Coincidentally, Curtin and Omak show up again, as in Chapter 4, although the high and low temperatures are reversed.

Figure 1 shows that, except for a market dip in the second half of 2015, this five-temperature model does a pretty good job matching the zigs and zags in stock prices. The model's accuracy is 0.60, which is very respectable for predicting something as unpredictable as stock prices.

How's that for knowledge discovery? Who knew that the daily high or low temperatures in these five small towns would be a useful predictor of stocks prices?

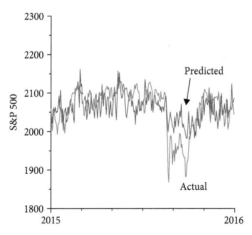

Figure 1 Some knowledge discovery for stock prices

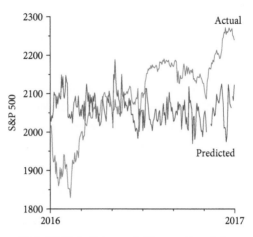

Figure 2 Maybe that was coincidence, not knowledge

The answer, of course, is that they are not. There is no plausible reason why the S&P 500 should be related positively or negatively to the high or low temperatures in these five cities, one of which is in Australia. We could make up a fanciful story about why daily stock prices depend on daily spending in these cities, and how spending depends on the temperature that day, but it would be utter nonsense.

This model was selected after estimating more than two million equations with 2015 data and picking the one equation that had the highest accuracy. Because it was based on data, rather than logic, we shouldn't expect it to work very well in predicting stock prices in 2016. Figure 2 shows that the accuracy in 2016 is –0.23. Yes, that is a negative sign. When the model predicted an uptick or downtick in stock prices, the opposite was likely to occur.

Try, try again

Disappointed by my temperature model, I considered 100 new variables and tried all possible combinations of one to five explanatory variables. The number of models was now near 80 million, but still small enough so that it was feasible for my data-mining software to try every possibility,

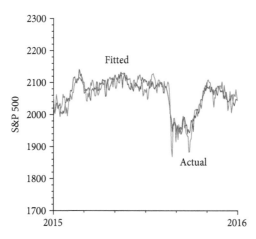

Figure 3 My five-variable model of stock prices

without having to resort to principle components, factor analysis, stepwise regression, or some other flawed data-reduction procedure.

The estimation of these models took several hours, so I stopped with five explanatory variables. If I had kept going, there are more than 17 trillion possible combinations of 10 variables, which would have taken my computer days to scrutinize. Fortunately, several five-variable combinations gave predicted stock prices that were closely correlated with actual prices. The best, shown in Figure 3, has an accuracy of 0.88. The fitted values are so close to the actual values that it is virtually impossible to tell them apart.

I may have unlocked the secret for predicting stock prices. Are you ready to invest?

I did this data-mining adventure in April 2017 using daily data for 2015. As with the five-temperature model, I had deliberately set aside daily data for 2016 in order to test my knowledge discovery. Figure 4 shows that the model that did spectacularly well in 2015 did spectacularly poorly in 2016, predicting stock price would collapse while they soared. Specifically, the model had an accuracy of 0.88 in 2015 and an accuracy of −0.52 in 2016. The five-variable model's predictions were strongly *negatively* correlated with the actual values of the S&P 500 in 2016. My model is worse than worthless.

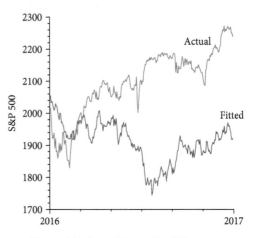

Figure 4 My five-variable model with fresh data

What happened? How can a model work so well one year and so badly the next? That is the nature of data mining. Choosing a model solely because it fits a given set of data really well virtually guarantees that it won't do nearly as well with fresh data. For a model to keep working with fresh data, it needs to make sense—and data-mining software cannot judge whether a model makes sense.

Here, I have tried to convince you of that reality by data-mining essentially random data—like the high temperature Curtin, Australia—that we all agree have essentially no effect on the S&P 500. Pushing that thought to its logical conclusion, the model in Figures 3 and 4 was not essentially random; it was completely random. The S&P 500 data are real, but I used a computer random number generator to create the 100 potential explanatory variables.

Remember how I once asked my students to create fictitious stock price data by flipping coins? Each stock's price started at $50 and then experienced a daily price change based on 25 coin flips, with the price going up 50 cents each time the coin landed heads and down 50 cents with each tail. I did this coin-flip experiment in class because I wanted my students to see for themselves how data that were clearly random could generate what appeared to be nonrandom patterns.

I did exactly the same thing here, this time using a computer's random number generator. I started each variable at 50 and used computerized coin flips to determine the daily change in the value of the variable. If the computer's coin flip was heads, the value went up 0.50; tails, it went down 0.50. I had the computer make 25 flips each imaginary day for each variable in order to create two years' worth of daily data for 100 imaginary variables. I labeled the first half of the random data 2015 and the second half 2016.

Even though all 100 variables were created by a random-walk process, it was nonetheless true, after the fact, that some variables turned out to be coincidentally correlated with the S&P 500. Among all five-variable possibilities, the combination of random variables 4, 34, 44, 64, and 90 was the most closely with the S&P 500 in 2015. When it came to 2016, the model flopped because these were literally random variables.

Black-box data mining cannot anticipate such a collapse because it cannot assess whether there is a logical basis for the models it discovers.

The set-aside solution

Now, it might be argued that, since the lack of any true relationship between the S&P 500 and my random variables was revealed by seeing how poorly the model fared in its 2016 predictions, we can use out-of-sample tests to distinguish between coincidental correlations and true causal relationships. Data mine part of the data for knowledge discovery and then validate the results by testing these discovered models with data that were temporarily set aside for this purpose. The original data are sometimes called *training data*, while the set-aside data are called *test data* or *validation data*. An alternative set of labels is *in-sample* (the data used to discover the model) and *out-of-sample* (the fresh data used to validate the model). In our S&P 500 examples using temperatures and random variables, the models were estimated using 2015 data and validated using the 2016 data that had been set aside for this purpose.

It is always a good idea to ask whether a model has been tested with uncontaminated data. It is never convincing to test a model with the very data that were plundered and pillaged to discover the model. It is certainly a good practice to set aside data to test models that have been tweaked or created from scratch to fit data.

However, tireless data mining guarantees that some models will fit both parts of the data remarkably well, even if none of the models make sense. Just as some models are certain to fit the original data, some, by luck alone, are certain to fit the set-aside data, too. Discovering a model that fits the original data and the set-aside data is just another form of data mining. Instead of discovering a model that fits half the data, we discover a model that fits all the data. That doesn't solve the problem. Models chosen to fit the data, either half the data or all the data, cannot be expected to fit other data nearly as well.

To demonstrate this, let's look at the 100 random variables that I created to explain fluctuations in the S&P 500. There are 100 one-variable models: Random Variable 1, Random Variable 2, and so on. In each case, I estimated the best fit using daily data for 2015. For Random Variable 1, this was:

$$Y = 2113.62 - 0.5489R1$$

The model's accuracy (the correlation between the predicted and actual values of the S&P 500) was 0.28. When I used this discovered model to predict the S&P 500 in 2016, the accuracy was −0.89. When the model predicted the S&P would go up, it usually went down, and vice versa.

I repeated this process of fitting the model to 2015 data and testing the model with 2016 data for all 100 potential explanatory variables. The results are shown in Figure 5. For the 2015 data used to estimate the model, the accuracy cannot be less than zero, because the model can always ignore the explanatory variable completely and thereby have an accuracy of zero. The average accuracy for the in-sample one-variable models was 0.27.

For the 2016 data that were set aside to test the models, the accuracy is equally likely to be positive or negative since these are, after all, random variables that have nothing at all to do with stock prices. We expect the average correlation between stock prices and any random variable to be around zero. For these particular data, the average out-of-sample accuracy happened to be -0.04.

Nonetheless, the out-of-sample accuracy will, by chance, be strongly positive for some models and strongly negative for others. The northeast corner of Figure 5 shows that there are several models that have reasonable accuracy for the 2015 in-sample data and also do well with the out-of-sample 2016 data. Specifically, eleven models have correlations above 0.5 using the 2015 fitted data, and five of these models have correlations

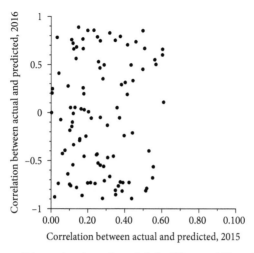

Figure 5 In-sample and out-of-sample fit for 100 one-variable models

above 0.5 using the set-aside data. These five models pass the out-of-sample validation test even though they are just random variables that are completely unrelated to stock prices.

With more explanatory variables, the accuracy should be higher. I repeated this exercise by estimating all 4,950 possible two-variable models. The best fit was with Random Variables 57 and 90:

$$Y = 2100.46 + 3.4612R57 - 4.8283R90$$

This model's in-sample accuracy was an impressive 0.79, but its out-of-sample accuracy was –0.56. The two-variable model that fit the data best in 2015 gave predictions in 2016 that were inversely related to the actual values. Despite this setback, additional data mining will inevitably turn up models that fit the 2015 training data well and the 2016 validation data, too.

With two possible explanatory variables, the average accuracy for the 2015 backtest data was 0.40. For the 2016 set-aside data, the average accuracy for the two-variable models was –0.01. Figure 6 shows the relationship between the 2015 accuracy and the 2016 accuracy.

The 4,950 models turn the graph into a giant blob. There are many models (like Random Variables 57 and 90) that fit the data reasonably well

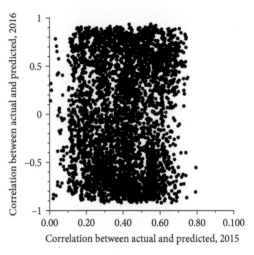

Figure 6 In-sample and out-of-sample fit for 4,950 two-variable models

in 2015 and do terribly in 2016, but there are also many models that fit the 2015 data well and also do well in 2016, sometimes even better than in 2015. That is the nature of chance, and these are chance variables.

There are 46 models with an accuracy above 0.70 in 2015, and 11 of these 46 models have an accuracy above 0.70 in 2016. These 11 models all pass the validation test even though it was pure luck. They are still useless for predicting stock prices in other years, like 2017.

One model, using Random Variables 14 and 74, had an accuracy of 0.70 for the 2015 data and an accuracy of 0.88 for the 2016 set-aside data! If we didn't know better, we might think that we discovered something important. But, of course, we didn't. All we really discovered is that it is always possible to find models that fit the in-sample and out-of-sample data well, even if the data are just random noise.

The same is, true, but even more so, for models with more explanatory variables. With additional variables, the number of possible models explodes, and so does the certainty of finding models that do well with fitted data and set-aside data. There are 161,700 possible three-variable models, 3,921,225 possible four-variable models, and 75,287,520 possible five-variable models.

As the number of possibilities grows, graphs like Figure 6 become a mass of solidly packed dots. But the principle still holds. There will inevitably be many models that fit the 2015 data well and also fit the set-aside data well.

For example, the best five-variable model had an accuracy of 0.88 in the 2015 in-sample data and an accuracy of –0.52 in the 2016 out-of-sample data. However, some five-variable models were lucky in 2015, some were lucky in 2016, and some were lucky in both 2015 and 2016. My data-mining software identified 11,201 five-variable models that had correlations of at least 0.85 between the predicted and actual values of the S&P 500 in 2015. Of these 11,201 models, 109 had accuracies above 0.85 in 2016, and 49 had accuracy above 0.90. If I had tried more variables, my data-mining software surely would have discovered some whose accuracy was above 0.90 and even 0.95 in both years.

That is not knowledge discovery. That is luck discovery.

If we ransack stock prices for a nonsensical system for beating the stock market, we will almost surely be poorer for it.

Real data mining

The Quantopian website allows investment-guru wannabes to write their own trading algorithms and backtest them on historical data to see how profitable the algorithms would have been. Sounds reasonable. Except we know that data mining will always yield algorithms that would have been profitable during the historical period used to discover the algorithms. And we know that the performance of algorithms that have little logical basis will usually be disappointing when used with fresh data, no matter how spectacular their backtest performance.

An interesting thing about the Quantopian platform is that although the details of the algorithms are not published, anyone can test them using any historical period they want. In addition, each algorithm is time-stamped to show the date when the final version was entered into the Quantopian system.

An outside group tested nearly a thousand stock-trading algorithms that were time-stamped between January 1, 2015 and June 30, 2015. Each algorithm was backtested using data for 2010 through the date of the time stamp (the training period), and then tested on fresh data from the time-stamp until December 31, 2015 (the test period). A comparison of the returns during the training and test periods found a small, but highly statistically significant *negative* relationship. Oops!

Convergence trades

When you short-sell a stock, you sell shares that you borrow from another investor. At some point, you have to buy the stock back (you hope at a lower price) and return them to the investor. Now, suppose that you could buy a stock for $90 and short-sell a virtually identical stock for $100. If the two prices converge to, say, $110, you have a $20 profit on the first stock (you bought it for $90 and sell it for $110) and a $10 loss on the second stock (you buy it for $110 after selling it for $100). Your net profit is $10, the difference between the two initial prices.

If, instead, the two prices converge to $80, you have a $10 loss on the first stock (you bought it for $90 and sell it for $80) and a $20 profit on the second stock (you buy it for $80 after selling it for $100). Your net profit is again $10.

This is called a *convergence trade* because you are not betting that the two stock prices will go up or down, but rather that the prices will converge to a common price.

Royal Dutch/Shell

The Royal Dutch Petroleum Company (based in the Netherlands) and the Shell Transport and Trading Company (based in the United Kingdom) joined forces in 1907 to do battle with John D. Rockefeller's Standard Oil, the largest oil refiner in the world. Royal Dutch would focus on production, Shell on distribution. Together, they might survive.

The curious thing about their agreement was that Royal Dutch and Shell kept their existing shareholders, and their two stocks continued to trade on various stock exchanges even though all revenue and expenses were consolidated by the mothership, named Royal Dutch Shell, which was owned 60 percent by Royal Dutch and 40 percent by Shell. Whatever earnings Royal Dutch Shell reported, 60 percent were attributed to Royal Dutch, 40 percent of Shell. Whatever dividends Royal Dutch Shell paid, 60 percent went to Royal Dutch shareholders, 40 percent of Shell shareholders. If Royal Dutch Shell were ever to be sold, 60 percent of the proceeds would go to Royal Dutch shareholders, 40 percent of Shell shareholders.

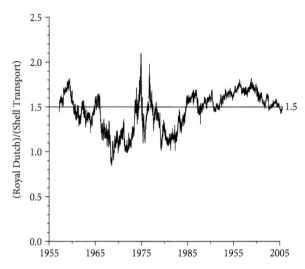

Figure 7 Royal Dutch Petroleum versus Shell Transport

Whatever Shell was worth, Royal Dutch was worth 50 percent more. If the stock market valued these stocks correctly, the market value of Royal Dutch stock should always be 50 percent higher than the market value of Shell stock. But it wasn't!

Figure 7 shows the ratio of the market value of Royal Dutch stock to the market value of Shell stock from March 13, 1957, when both stocks were first traded on the New York Stock Exchange, until July 19, 2005, when the two companies merged fully and their stocks stopped being traded separately.

The price of Royal Dutch stock was seldom exactly 50 percent more than Shell stock. Sometimes it was 40 percent overpriced; sometimes it was 30 percent underpriced. For the period as a whole, the percentage difference between the ratio of their market values and the theoretically correct 1.5 value was larger than 10 percent 46 percent of the time and larger than 20 percent 18 percent of the time.

This situation was ripe for a convergence trade. When Royal Dutch traded at a premium to the 1.5 ratio, an investor could have bought Shell and sold Royal Dutch stock short, betting on the premium disappearing.

This is exactly what one hedge fund, Long-Term Capital Management, did in 1997, when the premium was 8 to 10 percent. Long-Term bought $1.15 billion of Shell stock, sold $1.15 billion of Royal Dutch stock short, and waited for the market to correct. Long-Term was run by an all-star management team, including two finance professors who won Nobel prizes in 1997, and this was a smart bet because it was based on persuasive logic, not just a statistical pattern that may have been coincidental and meaningless. The ratio of the market values should go to 1.5 eventually and Long-Term would profit from its smart hedge.

However, as Keynes observed during the Great Depression:,

This long run is a misleading guide to current affairs. In the long run we are all dead. Economists set themselves too easy, too useless a task if in tempestuous seasons they can only tell us that when the storm is past the ocean is flat again.

Keynes was mocking the belief that, in the long run, the economy will be calm and everyone who wants a job will have a job. Keynes believed that the storm of an economic recession is more important than a hypothetical long run that no one will ever live to see. It is the same in the stock market. Convergence trades that might be profitable in the long run can be disastrous in the short run.

Long-Term's net worth at the beginning of 1998 was nearly $5 billion. In August, an unforeseen storm hit. Russia defaulted on its debt and perceived measures of risk rose throughout financial markets. Long-Term had placed bets in many different markets, but an awful lot of them were bets that risk premiums would decline. After the Russian default, risk premiums rose everywhere, and Long-Term was in trouble, big trouble.

Long-Term argued that all it needed was time for financial markets to return to normal—for the storm to pass and the ocean to become flat again—but it didn't have time. It had created tremendous leverage by making large bets with borrowed money—which is great if the bets pay off and catastrophic if they don't. Long-Term lost $550 million on August 21 and $2.1 billion for the entire month, nearly half its net worth.

Long-Term tried to raise more money so that it could wait out the storm, but frightened lenders didn't want to loan Long-Term more money. They wanted their money back.

Keynes was not only a master economist; he was also a legendary investor. He cautioned that, "Markets can remain irrational longer than you can remain solvent." Perhaps markets overreacted to the Russian default. Perhaps Long-Term's losses would have turned into profits eventually. But it couldn't stay solvent long enough to find out.

Long-Term had to close its Royal Dutch/Shell position at a time when the Royal Dutch premium, instead of declining, had gone above 20 percent. Long-Term lost $150 million on this trade.

On September 23, Warren Buffett faxed Long-Term a one-page letter offering to buy the firm for $250 million, roughly 5 percent of its value at the beginning of the year. The offer was take it or leave it, and would expire at 12:30 p.m., about an hour after the fax had been sent. The deadline passed and the funeral preparations began.

The Federal Reserve Bank of New York feared that the domino effects of a Long-Term default would trigger a global financial crisis. The Federal Reserve and Long-Term's creditors took over Long-Term and liquidated its assets. The creditors got their money back, Long-Term's founding partners lost $1.9 billion, and other investors got an expensive lesson in the power of leverage.

Notice in Figure 7 that the premium did disappear eventually, when the companies merged in 2005, with the Royal Dutch shareholders getting 60 percent of the shares in the combined company and Shell shareholders getting 40 percent. The Royal Dutch/Shell trade was a smart trade because it made sense and it did eventually pay off. Unfortunately, Long-Term made a lot of not-so-smart trades that forced it to liquidate the Royal Dutch/Shell trade prematurely.

The Royal Dutch/Shell mispricing is compelling evidence that stock market prices are sometimes wacky. Whatever the "correct" value of Shell stock, Royal Dutch was worth 50 percent more, yet stock market prices were sometimes higher, other times lower, than this, creating an opportunity for a profitable convergence trade. However, this example also demonstrates that convergence trades can be risky, even if done correctly by the best and brightest, because the convergence may take longer than expected. Convergence trades that have no logical basis are even riskier.

The GSR

In the 1980s, an investment advisory firm with the distinguished name Hume & Associates produced *The Superinvestor Files*, which were advertised nationally as sophisticated strategies that ordinary investors could use to reap extraordinary profits. Subscribers were mailed monthly pamphlets, each about 50 pages long and stylishly printed on thick paper, for $25 each plus $2.50 for shipping and handling.

In retrospect, it should have been obvious that if these strategies were as profitable as advertised, the company could have made more money by using their strategies than by selling pamphlets. However, gullible and greedy investors overlooked the obvious and, instead, hoped that the secret to becoming a millionaire could be purchased for $25, plus $2.50 for shipping and handling.

One Superinvestor strategy was based on the gold/silver ratio (GSR), which is the ratio of the price of an ounce of gold to the price of an ounce of silver. In 1985, the average price of gold was $317.26 and the average price of silver was $5.88, so that the GSR was $317.26/$5.88 = 54, which meant that an ounce of gold cost the same as 54 ounces of silver.

In 1986 Hume wrote that:

The [GSR] has fluctuated widely just in the past seven or eight years, dipping as low as 19-to-1 in 1980 and soaring as high as 52-to-1 in 1982 and 55-to-1 in 1985. But, as you can also clearly see, it has always—ALWAYS—returned to the range between 34-to-1 and 38-to-1.

Figure 8 confirms that the GSR fluctuated around the range 34 to 38 during the years 1970 through 1985.

Figure 8 The GSR 1970–1985

The GSR strategy is to sell gold and buy silver when the GSR is unusually high and do the opposite when the GSR is unusually low. The use of futures contracts to make these trades creates enormous leverage and the potential for astonishing profits. This is a convergence trade because the investor is not betting that gold or silver prices will go up or down, but that the ratio of their prices will converge to the historical ratio.

There is no logical reason why an ounce of gold should cost the same as 36 ounces of silver. Gold and silver are not like eggs sold by the dozen or half dozen, where, if the prices diverge, consumers will buy the cheaper eggs. Nor are gold and silver like corn and soybeans, where if the price of corn rises relative to the price of soybeans, farmers will plant more corn.

As it turned out, after the GSR went above 38 in 1983, it did not come back until 28 years later, in 2011. Figure 9 shows that the years when the GSR hovered around the range 34 to 38 were a temporary fluke, not the basis for a super strategy. Futures contracts multiply losses as well as gains and a 1985 bet on the GSR would have been disastrous.

Another convergence trade

Early convergence trades were based on simple patterns, like the ratio of gold prices to silver prices, that reveal themselves in price charts. Modern computers can ransack large data bases looking for more subtle and

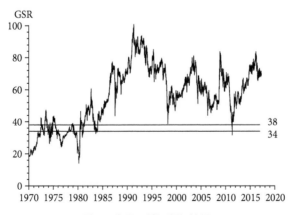

Figure 9 The GSR 1970–2017

complex convergence trades. If there has been a 0.9 correlation between two prices and the prices start to move in different directions, a trading algorithm might be tempted to wager that the historical relationship will reappear.

Even if a pattern is discovered by a computer, the problem is the same as with the GSR trade. Data without theory is treacherous. Convergence trades should make sense because, if there is no underlying reason for the discovered pattern, there is no reason for deviations from the pattern to self-correct. Statistical correlations may be coincidental patterns that appear temporarily and then vanish.

The Royal Dutch/Shell convergence trade had a logical basis, but Long-Term Capital Management was bankrupted by a number of bets on correlations that did not have a persuasive rationale; for example, risk premiums and relationships among various French and German interest rates. A manager later lamented that, "we had academics who came in with no trading experience and they started modeling away. Their trades might look good given the assumptions they made, but they often did not pass a simple smell test."

Ironically, looking at daily stock prices for Royal Dutch and Shell for any single year or two, a black-box trading algorithm would not have recognized that the ratio of their prices should be 1.5. It would have missed one of the few convergence trades that made sense.

Figure 10 shows a more recent convergence-trade opportunity. During 2015 and much of 2016, this ratio of two stock prices fluctuated around an average value of 0.76. The price ratio was sometimes well above 0.76 and sometimes well below, but it always returned to its apparent equilibrium.

On August 25, 2016, the price ratio poked above 1, suggesting it was a good time to sell one stock and buy the other. Alas, Figure 11 shows that the price ratio did not return to its natural equilibrium at 0.76, but instead went in the opposite direction, more than doubling—peaking at 2.14 on November 3, 2016, before retreating somewhat.

Perhaps the price ratio will go back to 0.76 someday. Maybe, but it won't be because special magnetic-like forces pull the price ratio towards 0.76.

How do I know this? Because I created these data using a random number generator. Once again, I used a computer's random number generator with the same rules as before. I started each stock's price at $50 and then used 25 computerized coin flips with 50-cent price swings to create two years' worth of daily stock price data for ten imaginary stocks.

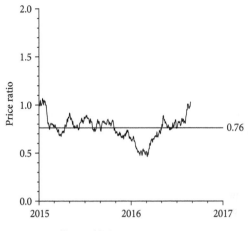

Figure 10 A convergence trade

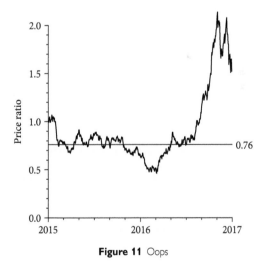

Figure 11 Oops

I then started looking at the ratios of pairs of prices. It didn't take long. The data in Figures 10 and 11 are the ratio of Random Price 2 to Random Price 1.

The ratio starts at 1 because each fictitious stock price starts at $50. By the luck of the flip, the ratio wanders around 0.76 for about a year and a half. Then it happens to spurt upward, before falling back. Where will the ratio go next? I don't know because it all depends on my computer's random number generator.

Random Price 1 and Random Price 2 are completely independent. They have nothing in common other than that they both start at $50. After that, every daily change in Price 1 was determined by 25 computerized coin flips and every daily change in Price 2 was determined by 25 different coin flips. Each price followed a random walk, equally likely to go up or down and fully independent of the other price path. And yet the ratio of their prices seemed to be tied to 0.76, never wandering away for long before coming back again. And then, suddenly, their random walks took the ratio far from 0.76, perhaps never to return.

The cautionary point is that a situation can appear ripe for a convergence trade even if the data are completely random. This doesn't mean that every potential convergence trade is random noise. What it does mean is that we can't tell whether a convergence pattern reflects something real or coincidental just by looking at the data, the way black-box data-mining software does. Computers have absolutely no way of judging whether a convergence pattern has a logical foundation. Only humans can decide if there is a persuasive reason for there to be a relationship. For Royal Dutch/Shell, the answer is yes. For the gold/silver ratio, the answer is no.

High-frequency trading

Some algorithms are used for high-frequency trading, buying and selling faster than humans can comprehend. A computer might notice that immediately after the number of stocks going down in price during the preceding 140 seconds exceeds the number going up by more than 8 percentage points, S&P 500 futures usually rise. The computer files this indicator away and waits. When this signal appears again, the computer pounces, buying thousands of S&P futures in an instant and then selling them an instant later. *Wired* marveled that these automated systems are "more efficient, faster, and smarter than any human." Faster, yes. Smarter, no.

Investment companies have spent billions of dollars building trading centers close to markets and using fiber optic cables, microwave towers, and laser links to shave milliseconds and nanoseconds off the time it takes to transmit information and trading orders between Chicago, New York, London, Frankfurt, and Tokyo. For example, a chain of microwave towers between the New York Stock Exchange and the Chicago Mercantile Exchange sends buy and sell orders more than 700 miles in less than 9 milliseconds round trip. For what purpose?

One purpose is to exploit perceived mispricings. Suppose that IBM stock can be bought for $200.0000 per share on one exchange and sold for $200.0001 on another exchange. A computer program that discovers this anomaly will buy as many shares as it can at $200.0000 for reselling a millisecond later for $200.0001 until the price discrepancy disappears. A profit of $0.0001 per share isn't much, but done hundreds or thousands of times in less than a second can generate a worthwhile annual return.

In a rational world, resources would not be wasted on such nonsense. Does it really matter if stock prices are ever so slightly different on different exchanges? Does it really matter if the mispricing lasts nine milliseconds rather than ten milliseconds?

A second purpose of lightning trades is to front-run orders entered by ordinary investors. If Jerry places an order to buy 1,000 shares of a stock at the current market price, a lightning trading program might buy the stock first and sell it milliseconds later to Jerry for a profit of a penny a share. A penny a share on 1,000 shares is $10. Repeated over and over, the profits turn into millions of dollars. What is the economic benefit to society of a computer ripping Jerry off by making him pay an extra penny a share? There is no benefit, just a computerized pickpocket stealing money without the victims even knowing that they were robbed.

More fundamentally, what is the economic benefit from having supremely intelligent programmers working on lightning trading programs when they could be doing something useful? What is the economic benefit of building trading centers and transmission lines to rush orders to markets when these resources could be used for something useful?

Even worse, lightning trades can cause economic harm.

If a human tells a computer to look for potentially profitable patterns—no matter whether the discovered pattern makes sense or not—and to buy or sell when the pattern reappears, the computer will do so—whether it makes sense or not. Indeed, some of the human minds behind the computers boast that they don't really understand why their computers decide to trade. After all, their computers are smarter than them, right? Instead of bragging, they should be praying.

The riskiness of high-frequency black-box investing is exacerbated by a copycat problem. If software engineers give hundreds of computers very similar instructions, then hundreds of computers may try to buy or sell the same thing at the same time, wildly destabilizing financial markets. To *Wired*'s credit, they recognized the danger of unsupervised computers moving in unison: "[A]t its worst, it is an inscrutable feedback loop . . . that can overwhelm the system it was built to navigate."

The flash crash

On May 6, 2010, the U.S. stock market was hit by what has come to be known as the "flash crash." Investors that day were nervous about the Greek debt crisis and an anxious mutual fund manager tried to hedge its portfolio by selling $4.1 billion in futures contracts. The idea was that if the market dropped, the losses on the fund's stock portfolio would be offset by profits on its futures contracts. This seemingly prudent transaction somehow triggered the computers. The computers bought many of the futures contracts the fund was selling, then sold them quickly because they don't like to hold positions for very long. Futures prices started falling and the computers decided to buy and sell more heavily. The computers were provoked into a trading frenzy as they bought and sold futures contracts among themselves, like a hot potato being tossed from hand to hand.

Nobody knows exactly what unleashed the computers. Remember, even the people behind the computers don't understand why their computers trade. In one 15-second interval, the computers traded 27,000 contracts among themselves, half the total trading volume, and ended up with a net purchase of only 200 contracts at the end of this 15-second madness. The trading frenzy spread to the regular stock market, and the flood of sell orders overwhelmed potential buyers. The Dow Jones industrial average fell nearly 600 points in 5 minutes. Market prices went haywire, yet the computers kept trading. Proctor & Gamble, a rock solid blue-chip company, dropped 37 percent in less than four minutes. Some computers paid more than $100,000 a share for Apple, Hewlett-Packard, and Sotheby's. Others sold Accenture and other major stocks for less than a penny a share. The computers had no common sense. They have no idea what Apple or

Accenture are worth. They blindly bought and sold because that's what their algorithms told them to do.

The madness ended when a built-in safeguard in the futures market suspended all trading for five seconds. Incredibly, this five-second time-out was enough to persuade the computers to stop their frenzied trading. Fifteen minutes later, markets were back to normal and the temporary 600-point drop in the Dow was just a nightmarish memory.

There have been other flash crashes since and there will most likely be more in the future. Oddly enough, Proctor & Gamble was hit again on August 30, 2013, on the New York Stock Exchange (NYSE) with a mini flash crash, so called because nothing special happened to other stocks on the NYSE and nothing special happened to Proctor & Gamble stock on other exchanges.

Inexplicably, nearly 200 trades on the NYSE, involving a total of about 250,000 shares of Procter & Gamble stock occurred within a one-second interval, triggering a 5 percent drop in price, from $77.50 to $73.61, and then a recovery less than a minute later. One lucky person happened to be in the right place at the right time, and bought 65,000 shares for an immediate $155,000 profit. Why did it happen? No one knows. Remember, humans aren't as smart as computers. Not.

Even if the exchanges implement safeguards to limit future flash crashes, these episodes illustrate the fundamental problem with black-box investing algorithms. These computer programs have no idea whether a stock (or any other investment) is truly cheap or expensive and do not even attempt to estimate what a stock is really worth. That's why a computer program might buy Apple for $100,000 and sell Accenture for less than a penny.

The bottom line

Computers don't have common sense or wisdom. They can identify statistical patterns, but they cannot gauge whether there is a logical basis for the discovered patterns. When a statistical correlation between gold and silver prices was discovered in the 1980s, how could a computer program possibly discern whether there was a sensible basis for this statistical correlation? When Procter & Gamble stock dropped 5 percent in an instant, how could a computer know whether this plunge was reasonable or ridiculous?

Humans do make mistakes, but humans also have the potential to recognize those mistakes and to avoid being seduced by patterns that lead computers astray.

One of my former students founded one of the first successful fund of funds, a strategy of investing in other investment funds, rather than buying stocks, bonds, and other assets directly. As part of his due diligence, he interviewed thousands of investors and fund managers. He identified four types of quant hedge funds (with some fund managers using a combination of strategies).

1 Pure arbitrage. Profit from trading identical or nearly identical assets, often at high frequencies. Examples include the same stock on two different exchanges or indices versus the underlying stocks. The profits are generally small but steady, with minimal risk.

2 Market makers. Exploit price discrepancies, such as similar securities traded at slightly different prices on different exchanges. Profits can be substantial, but there are risks that the trades will not be executed at the intended prices, particularly when exchanges are open at different hours or on different days (because of holidays).

3 Statistical arbitrage. Use data-mining algorithms to identify historical patterns that might be the basis for profitable trades. Profits can be large, but risks are also commensurately large. An example is buying stock in one airline while selling stock in another.

4 Fundamental quant. Use fundamental data like price/earnings ratios to favor stocks with certain attributes while avoiding or short-selling companies with opposite attributes.

His overall assessment of quants: "There are thousands of 'quant' investors and mutual funds. Only a handful have been able to deliver superior long-term results. Like musicians, there is not only one genre that is successful. A few Jazz, Rock and Country artists can sell out an entire arena while thousands of others play in cheap night clubs or on street corners."

CHAPTER 12

We're watching you

Humans often anthropomorphize by assuming that animals, trees, trains, and other non-human objects have human traits. In children's stories and fairy tales, for example, pigs build houses that wolves blow down and foxes talk to gingerbread men.

Think about these stories for a minute. The three little pigs have human characteristics reflected in the houses they build of straw, sticks, or bricks. The wolf uses various ruses to try to lure the pigs out of the brick house, but they outwit him and then put a cauldron of boiling water in the fireplace when they realize that the wolf is climbing up the roof in order to come down the chimney.

The gingerbread man is baked by a childless woman, but then runs away from the woman, her husband, and others, taunting his pursuers by shouting, "Run, run as fast as you can! You can't catch me. I'm the Gingerbread Man!" In some versions, a fox tricks the gingerbread man into riding on his head in order to cross a river and then eats him. In the version read to me when I was a child, a wily bobcat tries to lure the gingerbread man into his house for dinner, but birds in a nearby tree warn the gingerbread man that he *is* the dinner. The gingerbread man flees while the bobcat snarls, "Botheration!" The gingerbread man runs back home, where he is welcomed by his family and promises never to run away again.

These are enduring fairy tales because we are so willing, indeed eager, to assume that animals (and even cookies) have human emotions, ideas, and motives. In the same way, we assume that computers have emotions, ideas, and motives. They don't.

> Don't anthropomorphize computers.
> They hate that.

Nonetheless, we are fascinated and terrified by apocalyptic science-fiction scenarios in which robots have become smarter than us—so smart that they decide they must eliminate the one thing that might disable them: humans.

The success of movies such as *Terminator* and *Matrix* has convinced many that this is our future and it will be here soon. Even luminaries such as Stephen Hawking and Elon Musk have warned of robotic rebellions. In 2014, Musk told a group of MIT students, "With artificial intelligence we are summoning the demon.... In all those stories where there's the guy with the pentagram and the holy water, it's like yeah he's sure he can control the demon. Didn't work out." Three years later, Musk posted a photo of a poster with the ominous warning, "In the end, the machines will win." In 2017 Cambridge professor emeritus, Sir Martin Rees, predicted a "post-human" future in which machines could rule earth for billions of years. Fortunately, he thinks that the takeover will not happen "for a few centuries."

Every knowledgeable computer scientist I've talked to dismisses the idea of an imminent computer takeover as pure fantasy. Computers do not know what the world is, what humans are, or what survival means, let alone how to survive.

The more realistic danger is that humans will do what computers tell us to do, not for fear that computers will terminate us, but because we are awestruck by computers and trust them to make important—indeed, life and death—decisions. Remember how Hillary Clinton trusted Ada too much? She is hardly alone. Which mortgage applications will be approved? Which job applicants will be hired? Which people will be sent to prison? Which medicines should be taken? Which targets will be bombed? Too many people believe that, because computers are smarter than us, they should decide.

In many ways, I've been preparing you for this chapter. You will read things here that are astonishing—unless you take into account what we've discussed in earlier chapters.

A large part of what is meant by Big Data is Big Brother monitoring us incessantly. Big Brother is indeed watching, but it is Big Business as well as Big Government collecting detailed information about everything we do so that they can predict our actions and manipulate our behavior. Big Business and Big Government monitor our credit cards, checking accounts, computers, and telephones, watch us on surveillance cameras, and purchase data from firms dedicated to finding out everything they can about each and every one of us.

And they are watching—not only what we buy in stores or online, but which web sites we go to, how we use our cell phones, what cars we drive, what friends we have. Some businesses use their security cameras to track how we wander through their stores so that they can arrange their merchandise to entice us into buying more. Have you ever noticed that IKEA is laid out like a giant maze so that disoriented shoppers have to pass virtually every aisle to get to the section they want and, then, go past every aisle a second time in order to find their way back out of the store? This is no accident.

The avalanche of personal data collected by Big Business and Big Government is used to push and prod us to buy things we don't need, visit places we don't enjoy, and vote for candidates we shouldn't trust. Businesses and governments also use Big Data to decide who should be hired, fired, and imprisoned. Big Data knows best. Or not.

One obstacle is that humans are fickle, even irrational, so it is pretty hard to predict our behavior. Why did Beanie Babies and Cabbage Patch Kids become expensive collectibles, while other stuffed creatures came and went without a ripple? Why do hula-hoop crazes come and go? Why do some movies flop, while others break records?

William Goldman, a best-selling novelist who won two academy awards for his screenwriting, wrote in his memoir, *Adventures in the Screen Trade*:

The single most important fact, perhaps, of the entire movie industry:
  NOBODY KNOWS ANYTHING

Goldman gives two examples:

Raiders [of the Lost Ark] is the number four film in history as this is being written. . . . But did you know Raiders of the Lost Ark was offered to every single studio in town—and they all turned it down?
  All except Paramount.

Why did Paramount say yes? Because nobody knows anything. And why did all the other studios say no? Because nobody knows anything. And why did Universal, the mightiest studio of all, pass on Star Wars?…Because nobody, nobody—not now, not ever—knows the least goddamn thing about what is or isn't going to work at the box office.

We are impulsive, emotional, and susceptible to fads and follies. Predict that!

In addition, we've seen over and over again that data-mining algorithms will inevitably discover models that predict the past incredibly well, but are useless, or worse, for predicting the future. Big Data does not always know best, and it is perilous for businesses, governments, and ordinary citizens to make important decisions based on data-mined models—especially models hidden inside black boxes.

A pregnancy predictor

Target is the second largest discount retailer in the United States (Walmart is the largest). Target brands itself as more upscale than Walmart, catering to "the needs of its younger, image-conscious shoppers." Hence, the fake French pronunciation: *Tar-jay* or *Tarzhay*.

One area in which Target has excelled is the collection and analysis of customer data. Target assigns every customer a unique Guest ID number, which it uses to track everything every customer buys and every interaction with Target, including e-mails, web site visits, and coupons. To fill in its knowledge of its customers, Target collects and purchases additional data, including obviously valuable information (such as age, occupation, and estimated income) and less obviously useful data (such as taste in music, movies, and coffee machines).

In one interesting project, Target set out to identify pregnant customers. They figured that people are more likely to change their shopping habits during life-changing events like marriage, births, and divorce, and Target wanted to lure new mothers into the convenience of doing all their shopping at Target.

After mothers give birth, many stores use public birth records to bombard the family with baby-related coupons and offers. Target hoped that they could beat the rush by locking in expectant mothers *before* they give birth.

The task force assigned to this project did not use data-mining software to rummage through their vast data base, looking for statistical correlations, or, even worse, use data-reduction programs to transform their data into unrecognizable mish-mash. Instead, they built their models by identifying women who were in Target's baby registry and then analyzing changes in their buying habits. They also thought about how a human might guess that a woman is pregnant. They noticed that pregnant women tended to buy dietary supplements, like calcium, during the first half of their pregnancies, tended to switch to unscented soap and lotion during their second trimester, and stock up on cotton balls and washcloths during their third trimester. This all made sense to the humans monitoring the statistical analysis.

Target used data for 25 products to estimate the probability that a woman was pregnant and to predict her due date. Target applied its model to all of its customers and sent (ahem) targeted coupons and special offers to lure women who were likely to be pregnant into its stores. Target also used its data base to identify the kinds of offers—mailed flyers, e-mail discounts, or coupons for Starbucks coffee—that had been most successful historically for each individual customer.

The only problem was that the baby-related coupons alerted women to the fact that Target had somehow invaded their privacy and knew they were pregnant. Target had to modify its campaign by mixing baby-related offers with non-baby offers to make it less obvious that Target knew so much.

Target's model worked because it was managed by expert humans and not hidden inside a black box. Data-mined models are quite different. A data-mining approach would have applied a data-reduction procedure like principal components to all of the hundreds of characteristics of the women, even things that made no sense, like purchases of men's socks and cat food.

The results would have been an unintelligible goulash of apples, oranges, and potatoes. Target didn't do that, but other companies do.

Google Flu

In 2011, Google reported that its researchers had created an AI program called Google Flu that used Google search queries to predict flu outbreaks up to ten days before the Centers for Disease Control and Prevention

(CDC). They boasted that, "We can accurately estimate the current level of weekly influenza activity in each region of the United States, with a reporting lag of about one day." Their boast was premature.

Google's data-mining program looked at 50 million search queries and found the 45 key words that best fit 1,152 observations on the incidence of flu, using IP addresses to identify the states the searches were coming from. To reassure Google users that their privacy was not being violated, Google reported that Google Flu is a completely black-box program in that humans do not have any influence on the selection of the search terms, and that the selected key words are not made public. This revelation may have reassured people who were concerned about their privacy, but it should have been a red flag for people who know the weaknesses of black-box models.

In its 2011 report, Google said that its model was 97.5 percent accurate, meaning that the correlation between the model's predictions and actual CDC data was 0.975 during the in-sample period. An MIT professor praised the model: "This seems like a really clever way of using data that is created unintentionally by the users of Google to see patterns in the world that would otherwise be invisible. I think we are just scratching the surface of what's possible with collective intelligence."

However, after issuing its report, Google Flu over-estimated the number of flu cases for 100 of the next 108 weeks, by an average of nearly 100 percent. Out-of-sample, Google Flu was far less accurate than a simple model that predicted that the number of flu cases in the coming week will be the same as the number last week, two weeks ago, or even three weeks ago. Since cases of the flu rise in the winter and fall in the summer, a simple model based on daily temperatures would have done better than Google's black box.

Google said that one possible explanation for Google Flu's out-of-sample flop was that it may have used generic search terms like "fever" which are not necessarily flu-related. Or perhaps people thought they had the flu, but didn't. Ordinary people aren't doctors, after all.

All true, but the data were just as noisy during the training period and its accuracy then was 97.5 percent. Garbage in, Garbage out. Why would anyone think that garbage turns into gold just because it has Google's name on it?

Cynics suggested that, since flu outbreaks are highly seasonal, Google Flu may have been mostly a winter detector that picked out seasonal search terms like New Year's Eve and Valentine's Day. Remember, how the high temperature in Omak, Washington, predicted the low temperature the next day in Curtin, Australia, even though there was no causal relationship between the two? The same thing probably happened here. With 50 million search terms, there are bound to be some that are coincidentally correlated with flu outbreaks, birth rates, and the price of tea in China. It works great, until it doesn't.

Google Flu no longer makes flu predictions.

Robo-Tester

An analytics firm (I'll call them WhatWorks) tests different web-page designs for internet companies. WhatWorks bases its recommendations on experiments in which a computerized random-event generator sends customers to pages with different web designs. After several days of tests, WhatWorks tells the client which web design generated the most revenue per visitor.

WhatWorks was hired by an internet company (BuyNow) that manages over a million domains. WhatWorks did its tests and reported back an optimal design. The client had a better idea. Visitors to a different domains might have different preferences, so why not allow for different designs for each domain? BuyNow did not know how visitors might differ or how these differences might affect their preferences, but surely customized designs were better than one design fits all.

This couldn't be done manually with human experts making decisions for individual domains, because it would take more than a million experiments. Also, many domains had very little web traffic and it would take months to reach a decision. So, why not harness the power of machine learning? BuyNow paid WhatWorks to develop Robo-Tester, an automated AI system that would use a variety of designs for each web site, monitor which designs worked best, and as its confidence grew, increase the frequency with which the favored designs are used.

Good in theory, flawed in practice. BuyNow thought this AI system would provide definitive answers, but there is a lot of random noise in web traffic and in the revenue generated by web traffic. An inferior design might, by luck alone, look good for a while, tricking Robo-Tester into favoring this inferior design. In addition, the random noise meant that

Robo-Tester was never 100-percent sure that it had found the best design, so it kept testing loser designs.

The net result was that Robo-Tester reduced BuyNow's revenue by about 1 percent, compared to the single design initially recommended by WhatWorks' experts.

The client complained, even though it was the client's idea:

BuyNow: *What's wrong with using AI to automate this on every domain?*
WhatWorks: *Your domains don't have enough traffic to draw meaningful conclusions at an individual level.*
BuyNow: *But what about only switching layouts when there is statistical evidence?*
WhatWorks: *Statistical evidence doesn't mean it's significant.*
(stunned silence)
WhatWorks: *I mean, with a million tests you'll get statistical evidence all over the place, but that doesn't mean it's picking the best design.*
BuyNow: *Yeah, you always say that.*
WhatWorks: *And I'm always right!*

WhatWorks *was* right, but people don't want to hear that AI may be counterproductive. BuyNow terminated its contract with WhatWorks.

Job applications

It is really, really hard to evaluate job applicants based on quantifiable data like grade point average and years of job experience. Even at Google, the temple of Big Data, Lazlo Bock, Google's Vice President for People Operations, was brutally candid in 2016:

"Tell me about yourself. What's your greatest strength, greatest weakness?" None of those common interview questions tell you anything about whether the person is going to be good in the job.

The only questions that do are questions that actually say, "Give me an example of work you did that is exactly like the work you're going to do." They're called "structured behavioral interview questions." Everything else is kind of a waste of time.

How could an AI program conduct and evaluate structured behavioral interview questions?

For people who have not done exactly the work they are being hired to do, the most important quality is an ability to learn to do the job quickly and well. A great hire will understand what information is needed, what to do with that information, and how to deal with surprises and ambiguity.

AI programs cannot gauge any of that so, instead, they might screen job applications by looking for specific keywords, like *Excel, sales, mathematics*, and *PhD*. AI algorithms have no flexibility and it is inconceivable that keywords are the secret to evaluating job candidates. Unless the people who created the algorithm build it into the program, an AI algorithm looking for *computer science* may overlook *data analytics*, looking for *PhD* may overlook *Ph.D.*, and looking for *mathematics* may overlook *math*. An applicant who is fluent in Excel may not think that is worth putting in a résumé. An AI algorithm won't know the difference between being founder of *MakeMyDay*, an internet gift company, and *MakeMyDay*, a club devoted to watching Clint Eastwood movies. An AI algorithm may not know the difference between a 3.2 GPA at CalTech (an average genius) and a 3.2 GPA at schools I won't name that are easy to get into and where grade inflation has raised the average grade to an *A*.

Conversely, job applicants can game the system by including the requisite key words in their cover letters and résumés; for example, stating an interest in *mathematics* instead of a degree in mathematics may fool an AI algorithm.

If you have ever seen successful and unsuccessful hires, you know that the difference between the hires that worked out well and those that disappointed had little or nothing to do with key words in résumés and job applications.

Some companies make data-driven software that evaluates job applicants by monitoring their online activities. The chief scientist for one company acknowledged that some of the factors chosen by its AI data-mining software do not make sense. For example, the software found that several good programmers in its data base visited a particular Japanese manga site frequently, so it decided that people who visit this site are likely to be good programmers. The chief scientist said that, "Obviously, it's not a causal relationship," but argued that it was still useful because there was a strong statistical correlation. Ouch! This is just another example of an ill-founded belief—even by people who should know better—that data are more important than common sense. They are not.

She also said that the company's algorithm looks at dozens of variables, and constantly changes the variables as correlations come and go. She believes that the ever-changing list of variables demonstrates the model's power and flexibility. A more compelling interpretation is that the algorithm captures transitory coincidental correlations that are of little value. If these were causal relationships, they would not come and go. They would persist and be useful. It is just like a technical analysis of stock prices that uncovers fleeting patterns. Make up model. Test. Change model. Repeat.

Was this firm's software successful in identifying good job candidates? A person who worked for the company for three years wrote that:

Customers really, really hate the product. There are almost no customers that have an overall positive experience. This has been true for years, and management is not able to reimagine the company in a way that would let them fix that core problem. The product team doesn't do any product research or incorporate feedback into their designs in a meaningful or organized way, so this will never get fixed.

Job advertisements

Computer algorithms are also used for job recruitment. The Introduction to this book discussed how politicians use microtargeting to make carefully crafted pitches to narrowly defined groups of potential voters. Businesses do the same; in fact, they were the first to do it—using large data bases to identify consumers thought to be especially receptive to specific pitches for particular products.

Microtargeting gives internet companies with massive amounts of data about individuals a huge advantage over newspapers, magazines, radio, television, and other traditional advertising media. Instead of buying ads that are read, heard, or seen by mass audiences, businesses can focus on potential customers.

The same is true of job advertisements. Instead of blasting a job opening to a mass audience, businesses can target potential job applicants who might be qualified and interested in the job. Unfortunately, this can result in intentional or unintentional discrimination.

A 2017 study by *The New York Times* and ProPublica, a non-profit online news service that won a Pulitzer Prize for its investigative journalism, found that microtargeted job ads frequently discriminate against older workers. One example was a Verizon ad for a finance position that was sent to Facebook users who were interested in finance, lived in or had recently visited Washington, D.C., and were between the ages of 25 and 36. A United Parcel Service job ad targeted people 18 to

24; a State Farm ad targeted ages 19 to 35. Facebook, itself, targeted' ages 25 to 60.

Some companies argue that this targeting isn't any different from placing ads in magazines with young readers or on radio stations with youthful listeners. There is a reason that daytime television has lots of targeted ads for vocational training schools and personal injury lawyers. However, speaking as someone over the age of 60, I can read whatever magazines I want, listen to whatever radio stations I want, and watch whatever television I want. I can't see Facebook ads that are not sent to me. I don't even know they exist.

Job-application algorithms may also discriminate in unintended ways. A 2016 experiment evaluated the effectiveness of a social media ad that was intended to be a gender-neutral vehicle for recruiting people for Science, Technology, Engineering and Math (STEM) jobs. Despite the explicit gender-neutral intention, the number of men who saw the ad was 20 percent larger than the number of women who saw the ad.

The researchers thought that perhaps women were less likely to click on the ad, which would lead an AI algorithm that was trying to maximize ad clicks to move away from web sites with higher female traffic. No, women were more likely to click on the ad. The reason why fewer women saw the ad was more subtle and far more likely to be detected by a human than by a computer algorithm.

Women in the 25–34 age bracket are valuable internet customers because they are more likely to click on ads selling products, and more to likely purchase products. Ads are consequently more expensive on internet sites with relatively high female traffic, and these high prices are charged for ads for jobs as well as products. An AI algorithm that is designed to place job ads while minimizing costs will consequently veer away from relatively expensive sites that are visited more frequently by females.

The researchers thought of this creative answer and were able to test it in a variety of ways. An AI algorithm would have been clueless.

Loan applications

Loan applications have traditionally been assessed by using sensible measures of an applicant's ability to repay the loan and history of repaying loans and other bills. Repay ability depends on employment history, income, wealth, and other debts. Bill-payment history is simply a record of how many payments have been late or were never paid. The 5 Cs of lending are:

Character:	Does the applicant have a history of paying bills on time? A criminal record?
Capacity:	Does the borrower have a stable employment history and more than enough income to make these loan payments in addition to other obligations?
Capital:	Does the borrower have sufficient wealth to make the loan payments if income drops?
Conditions:	Is the borrower working in a volatile industry?
Collateral:	Is there sufficient collateral for the lender to get its money back if the borrower defaults?

With the explosion of the internet and social media, lenders now look for other, less direct evidence, particularly of a person's character. For example, going to respectable web sites and interacting with respectable people makes you look respectable. Yes, it's like the dating game in school, with the popular and unpopular groups. And, yes, you can game the system. Don't say anything embarrassing or publicize your embarrassing behavior. Read serious internet articles written by serious people. Follow serious people. Post serious comments on serious web sites.

It was only a matter of time before lenders realized that they could move beyond manually assessing the 5 Cs and using humans to search the internet for a loan applicant's name to see if anything embarrassing pops up. They could tap the power of AI to evaluate loan applicants!

In 2017, the founder and CEO of a Chinese company making an AI lending app argued that,

While banks only focus on the tip of the iceberg above the sea, we build algorithms to make sense of the vast amount of data under the sea.

What useful data are under the sea? You might be surprised to learn that it is all about the smartphones. Many Chinese citizens are paid in cash, buy things with cash, do not have credit cards, and have never borrowed money from banks. But the majority have smartphones, and many get small short-term loans from online lenders.

This Chinese company's data-mining software has access to literally hundreds of millions of loan applications made through smartphones and, in addition, enough information about the smartphones themselves for AI programs to detect statistical patterns. The CEO bragged that,

We don't hire any risk-control people from traditional financial institutions. . . We don't need human beings to tell us who's a good customer and who's bad. Technology is our risk control.

Among the data that show up as evidence of a person being a good credit risk: using an Android phone instead of an iPhone; not answering incoming calls; having outgoing calls not answered, and not keeping the phone fully charged.

We could invent theories to explain the discovered statistical patterns. People who use inexpensive Android phones are financially prudent. People who do not always answer incoming calls are serious workers. People whose outgoing calls are not always answered have friends who are equally productive. People who do not keep their phones fully charged are more focused on their work than on their phones. Are you persuaded?

If so, you might be surprised to learn that the results were actually the opposite. These are the indicators of being a bad credit risk! Of course, we could make up stories for that, too. People who use Android phones can't afford iPhones. People who don't answer incoming calls are avoiding creditors. People whose outgoing calls go unanswered are disliked. People who let their phone batteries run down are irresponsible.

The point is that humans are clever enough to invent plausible stories for whatever statistical patterns are discovered, even if the statistical patterns are random noise discovered by data-mining software. Finding patterns proves nothing. Making up stories to fit the patterns proves nothing.

I predict that, no matter how well the model fits the data, these algorithms are going be poor predictors of loan defaults.

Car insurance

In 2016, Admiral Insurance, Britain's largest car insurance company, planned to launch "firstcarquote", which would base its car insurance rates on an AI analysis of an applicant's Facebook posts:

We already know social media posts can tell us whether a person is a good or a bad credit risk and this is true for cars too. It's scientifically proven that some personalities are more likely to have an accident than others. But standard insurance questions don't tend

to measure personality. At firstcarquote, we look at a driver's personality by analysing some of their Facebook data and if we see indicators that you will be a careful driver, we will give you a discount of between 5 and 15 percent off the price you would get on admiral.com.

Admiral executives said that the AI algorithm would look for indications in Facebook posts that an applicant is careful and cautious, as evidenced by making lists and setting specific times to meet, instead of "tonight." The Admiral advisor who designed the algorithm offered a semi-plausible justification:

Overconfident people might use words like 'always' or 'never'. Uneasy drivers are likely to use more negative emotions, so more words like 'maybe' or 'perhaps', which suggest that they are so not confident. . . . You can infer a few things about personality, and from the personality we can conclude how safe you're likely to be.

This explanation is only semi-plausible, however. Word choices might tell us something about an applicant's personality, but it is hard to see how this translates into a reliable predictor of whether a person is accident prone.

Then, the Admiral advisor's explanation wandered off into even more tenuous territory, claiming that liking Michael Jordan or Leonard Cohen might be a good predictor of car-insurance claims. Finally, he inadvertently admitted the truth in what was intended to be a boast:

Our algorithm for calculating what 'safe' looks like is constantly learning, as we match social data to actual claims data. . . . Our analysis is not based on any one specific model, but rather on thousands of different combinations of likes, words and phrases and is constantly changing with new evidence that we obtain from the data. As such our calculations reflect how drivers generally behave on social media, and how predictive that is, as opposed to fixed assumptions about what a safe driver may look like.

Well, there it is. No "fixed assumptions" means no logical basis. The algorithm is just a black box data-mining model that looks for historical correlations, with scant regard for whether they make sense, and the algorithm changes constantly because it has no logical basis and is consequently buffeted by short-lived correlations. Make up model. Test. Change model. Repeat.

We never had the opportunity to see how this algorithm would have fared because, a few hours before the scheduled launch, Facebook said that it would not allow Admiral to access Facebook data, citing its policy that "prohibits the use of data obtained from Facebook to make decisions about eligibility, including whether to approve or reject an application or how much interest to charge on a loan."

Probably a blessing in disguise for Admiral.

Social credit scores

Chinese business currently cooperate with the government in collecting data that track what people buy, where they go, what they do, and anything else that might suggest that a person is untrustworthy—not just less likely to repay a loan, but also more likely to foment political unrest.

Obedient citizens who are assigned high scores by the computer algorithm receive price discounts, pay low insurance rates, and can rent apartments without making deposits. Those given low scores are not permitted to buy certain things and must pay more for what they are allowed to buy. Their travel options and living arrangements are limited and they may be monitored by the police. One company's chief executive said that these scores "ensure that the bad people in society don't have a place to go, while good people can move freely and without obstruction."

Some of the data behind these scores are sensible; for example, using loan-payment history to predict loan defaults. However, some data are surely just temporary correlations; for example, predicting loan defaults from video games played and movies watched. Some monitoring—whether one spreads rumors or criticizes the government—is intended to curtail free speech.

A person's score is affected negatively by having friends with low scores. If Alice's score drops, some friends will abandon her lest their scores drop too—which lowers Alice's score even more. This guilt-by-association divides society since the haves are rewarded for keeping the have-nots out of their social networks.

Don't think China is alone in using algorithms and social media data to assess people. Facebook was not being altruistic when it blocked Admiral's use of Facebook data to price insurance. Facebook has its own patented algorithm for evaluating loan applications based on the characteristics of you and your Facebook friends. As I write this, Facebook has not yet implemented its algorithm, but don't think that they are not considering this and other ways to monetize their data.

Black-box discrimination

Job advertisements that are microtargeted are a clear example of a deliberate attempt to favor certain groups (and, inevitably, exclude others). However, such inequities are hidden if the microtargeting is done with a black-box algorithm that may discover statistical patterns that implicitly discriminate based on age, gender, or race.

No one, not even the programmers who write the code, know exactly how black-box algorithms make their predictions, but it is almost certain that employment, loan, and insurance algorithms directly or indirectly consider gender, race, ethnicity, sexual orientation, and the like. It is not moral or ethical to penalize individuals because they share characteristics of groups that a black-box algorithm has chosen as predictive of bad behavior.

Consider that algorithm that evaluates job candidates by looking at the websites that applicants visit—not "good" or "bad" websites, just sites that may or may not be correlated with the sites visited by current employees. How fair is it if a Hispanic female does not spend time at a Japanese manga site that is popular with white male software engineers?

Consider that algorithm that evaluates loan applications by considering how frequently incoming cellphone calls are answered. How fair is it if a Jewish male is more or less likely to answer calls from telemarketers?

Consider that algorithm that sets car insurance rates based on Facebook word choices and whether a person likes Michael Jordan or Leonard Cohen. How fair is it if a black male likes Michael Jordan and a white female likes Leonard Cohen? How fair is it if word choices that are related to gender, race, ethnicity, or sexual orientation happen to be coincidentally correlated with car insurance claims?

Unreasonable searches

It has been argued that data-mining algorithms applied to "seemingly unrelated data" can "discover, deter, and detect crime." Oh, boy, fighting crime with admittedly irrelevant data.

For example, maybe we can use data-mining algorithms instead of courts to authorize search warrants? We seem headed down that road, but it won't be a pleasant journey.

Search warrants are generally limited in scope and must be based on specific, individualized facts that do not have an obvious innocent explanation. There is no obvious innocent explanation if a drug-sniffing dog selects a piece of luggage, if a voice is heard screaming from inside a car trunk, or if drug-making equipment is seen being carried into a basement. There is an innocent explanation if a family locks their front door (they are afraid of burglars) or flies to Colombia (they may have family there or be going for a vacation).

Here is a more complex example. In Illinois, the police obtained a search warrant for a tavern based on evidence that the tavern owner was selling heroin. When police entered the tavern, they searched the customers and found that Ventura Ybarra had heroin hidden in a cigarette pack. The Supreme Court ruled that the search of Mr. Ybarra was illegal because it was based on guilt by association, rather than any specific probable cause. The police did not know Ybarra, who had been playing pinball, and he did not attempt to flee, hide something, or exhibit any other suspicious behavior. There was an obvious innocent explanation for his presence in the tavern: he had come inside to have a drink and play pinball.

The Court stated that "probable cause [must be] particularized . . . to [each] person," a standard that can never be satisfied by "pointing to the fact that coincidentally there exists probable cause to search [another person] or to search the premises where the person may happen to be."

This is perhaps an insurmountable problem for using a data-mining algorithm to issue search warrants. Computers are very inept at imagining innocent explanations for very specific—indeed unique—situations. Remember the cat-and-vase example in Chapter 3? Without being able to draw on human life experiences, computers cannot imagine plausible explanations in context. Humans have seen the wind blow things over; so we consider that possibility. Humans have seen children knock things over when they run through a house; so we consider that possibility. Humans have felt earthquakes in the area; so we consider that possibility. Computer software can do none of this. The same is true of search warrants. AI algorithms cannot imagine, let alone assess, innocent explanations.

A related issue is that a search warrant should have a plausible basis. It is plausible that illegal drugs might be manufactured in an abandoned building. It is less plausible that illegal drugs would be manufactured in a church or senior-citizens center. A data-mining algorithm would not know the difference.

Another problem with data-mining algorithms is that they may use proxies for race, religion, or political associations that should not be considered in search-warrant applications. For example, an algorithm rummaging through Big Data might find that living in certain areas, purchasing certain products, wearing certain clothing, or visiting certain web sites are good predictors of criminal activity but are also proxies for race, religion, and other protected classes.

One reason that the courts require specific, individualized evidence of criminal activity is that vagueness fosters government abuse. First, citizens should know what specific activities are illegal or can be the basis for invasions of their privacy. Second, the government should have limited discretion in deciding whose privacy should be invaded.

Suppose that a data-mining algorithm issues a search warrant or arrest warrant based on activities that are legal, but statistically correlated with illegal activities. Most of the patterns picked up are likely to be temporary random noise. Remember that the bigger the data set being ransacked, the more likely it is that a discovered pattern will be a meaningless coincidence.

What if an algorithm discovers a correlation between illegal drug use and buying dog food at Costco, watching Seinfeld, and buying a new car every three years? These may be completely coincidental statistical relationships, like the correlation between presidential election outcomes and the temperatures in five small towns. Even if it isn't completely coincidental, shouldn't citizens be forewarned that they might be arrested or have their homes searched if they watch Seinfeld reruns?

As for the second issue—police discretion leading to selective law enforcement—imagine that a data-mining algorithm identifies 10,000 people who have characteristics similar to known criminals. If the police do not have the resources for a thorough search of all 10,000 residences, they may choose a few dozen residences that they consider especially promising, thereby undermining the principle that searches be based on evidence, not whim.

Watch your wristbands

Richard Berk has appointments in the Department of Criminology and the Department of Statistics at the University of Pennsylvania. One of his specialties is algorithmic criminology: "forecasts of criminal behavior and/or victimization using statistical/machine learning procedures." Algorithmic criminology is becoming increasingly common in pre-trial bail determination, post-trial sentencing, and post-conviction parole decisions.

Berk writes that, "The approach is 'black box,' for which no apologies are made." In an article in *The Atlantic*, Berk is more explicit: "If I could use sun spots or shoe size or the size of the wristband on their wrist, I would. If I give the algorithm enough predictors to get it started, it finds things that you wouldn't anticipate."

Most "things that you wouldn't anticipate" are things that don't make sense, like sun spots, shoe sizes, and wristband sizes. They reflect temporary, coincidental patterns that are useless predictors of criminal behavior. A study of one of the most popular risk-assessment algorithms found that only 20 percent of the people predicted to commit violent crimes within two years actually did so—and that the predictions discriminate against black defendants.

Berk no doubt has good intentions, but it is unsettling that he thinks people should be paroled or remain incarcerated based on sunspots, shoes, and wristbands. That's what happens when you trust computers too much.

Even worse, the companies selling criminology algorithms generally do not disclose what factors they look at and which factors their algorithms decide are important, arguing that this is valuable proprietary information. This secrecy makes it impossible for defendants to defend themselves from being incarcerated based on coincidental patterns in the data.

In one case, Wisconsin police identified a car involved in a drive-by shooting and chased after it. When the car stopped, the driver, Eric Loomis, was arrested and soon pleaded guilty to eluding a police officer and pleaded no contest to driving another person's car without the owner's consent. Based in part on a computer algorithm's prediction that Loomis was likely to commit more crimes if he was not incarcerated, the judge sentenced him to six years in prison. Loomis' appeals to the Wisconsin Supreme Court and the U.S. Supreme Court were rejected.

The Wisconsin Supreme Court lauded evidence-based sentencing—using data instead of subjective human opinion. Human opinion is certainly imperfect, but so are black-box data-mining algorithms. The Court argued that although Loomis could not challenge how the "algorithm calculates risk, he can at least review and challenge the resulting risk scores." It is hard to see how he could do that, other than by creating his own algorithm.

In the Supreme Court Appeal, the Wisconsin Attorney General argued that the Court should adopt a wait-and-see attitude: "The use of risk assessments by sentencing courts is a novel issue, which needs time for further percolation." In the meantime, Loomis was "was free to question the assessment and explain its possible flaws." Again, realistically, how can he challenge a black-box model?

The makers of one popular algorithm did reveal a little about their inputs, but it was not comforting. Among the factors they considered were the defendants' answers to such questions as, "How many of your friends/acquaintances are taking drugs illegally?" and "If people make me angry or lose my temper, I can be dangerous." Seriously, who would answer, "Dozens" and "Certainly"?

Risk-assessment algorithms surely ignore the answers to questions that any sensible criminal would lie about. They must be based on more subtle data (like wristbands and shoe sizes), and it certainly seems fair that defendants should know what those data are, instead of rotting in prison while algorithms percolate.

Do you need plastic surgery?

If parole is based on statistical models predicting recidivist behavior, it is just a short step to using statistical models to decide who should be arrested and imprisoned. Sure enough, in 2016, two Chinese researchers reported that they could predict with 89.5 percent accuracy whether a person is a criminal by applying their AI algorithm to scanned facial photos.

My first thought—and perhaps yours, too—was that this was a hoax, like the dead salmon study in Chapter 6. Perhaps two pranksters were giving a dramatic demonstration of how data mining can be used to concoct farfetched claims. With hundreds of thousands of pixels, there will inevitably be close correlations to *something*. The model could just as well create a phony predictor of a person's IQ, movie preferences, or susceptibility to cancer by scanning faces and using data-mining software to find statistical relationships.

I read the article and was surprised to find that, as far as I can tell, they are completely serious. They boasted that,

They scanned 1,856 male ID photos—730 criminals and 1,126 non-criminals—and their AI program found "some discriminating structural features for predicting criminality, such as lip curvature, eye inner corner distance, and the so-called nose-mouth angle."

MIT Technology Review was optimistic: "All this heralds a new era of 'anthropometry, criminal or otherwise,' and there is room for more research as machines become more capable." A respected data scientist wrote that, "the study has been conducted with rigor. The results are what they are." Spoken like a true data miner. Who needs theories? If an AI program finds statistical patterns, that's proof enough. The results are what they are.

To their great credit, AI researchers, as a whole, dismissed this study as perilous pseudoscience—unreliable and misleading, with potentially dangerous consequences if taken seriously.

The authors' published article includes photos of three criminal faces and three non-criminal faces, no doubt chosen so that readers would agree that criminals and non-criminals *do* look different. The criminals are not smiling, are not wearing business suits, and have rougher skin. Beyond lip curvature, eye inner corner distance, and the so-called nose-mouth angle (whatever these mean), is the AI program detecting the same things we notice? Is it anything more than a smile detector?

I still suspect that this study may be a hoax but, if it is not a hoax, then what, exactly, is the purpose of an AI criminal-face algorithm? Are we supposed to arrest people with criminal faces so that they cannot commit crimes?

I remember *Minority Report*, a 2002 movie in which three psychics ("precogs") are able to visualize murders before they occur, thereby allow-ing special PreCrime police to arrest would-be assailants before they can commit the murders. The logical problem, of course, is that if would-be murderers are arrested before they kill anyone, then the murders don't occur. So, how can precogs visualize murders that don't happen?

The criminal-face AI algorithm doesn't have this time-paradox, because it does not predict specific crimes. It just identifies people who the program calculates are 89.5 percent likely to be criminals. Their crimes may have been committed in the past.

Still, is the AI program to be used to arrest and imprison people for crimes that they got away with and/or pre-criminals before they commit crimes?

A blogger commented on the research:

What if they just placed the people that look like criminals into an internment camp? What harm would that do? They would just have to stay there until they went through an extensive rehabilitation program. Even if some went that were innocent; how could this adversely affect them in the long run?

I hope that this person was being sarcastic, but I fear that he or she wasn't. If this blind faith in computers becomes the norm, governments may well start imprisoning people based on an AI analysis of their faces.

What if you have lip curvature, eye inner corner distance, and the so-called nose-mouth angle? Criminal or not, plastic surgery may be in your future.

Gaming the system

There are two fundamental problems with data-mining algorithms. If an algorithm is a proprietary secret, we have no way of checking the accuracy of the data used by the algorithm. If a black-box algorithm is told that you defaulted on a loan, when it was really someone with a similar name, you won't know that there has been a mistake or be able to correct it. On the other hand, if the algorithm is public knowledge, people can game the system and thereby undermine the validity of the model. If an algorithm finds that certain word usage is common among people who have defaulted on loans, people will stop using those words, thereby getting their loans approved no matter what their chances of defaulting.

People who want low car insurance rates can make lists on Facebook. Prisoners who want to be paroled can change their wrist bands. Criminals who don't want to be arrested can have plastic surgery. Once people learn the system, they can game it—which undermines the system. This gaming phenomenon is so commonplace that it even has a name. Goodhart's law (named after the British economist Charles Goodhart) states that, "When a measure becomes a target, it ceases to be a good measure." Goodhart was an economic advisor to the Bank of England and his argument was that setting monetary targets causes people to change their behavior in ways that undermine the usefulness of the targets. We now know that Goodhart's law applies in many other situations as well.

When the Soviet Union's central planning bureau told nail factories to produce a certain number of nails, the firms reduced their costs by producing the smallest nails possible—which were also the least useful. Setting the target undermines the usefulness of the target.

Schools throughout the United States now administer standardized tests that are used to assess students, teachers, and schools. Teachers who do well may get bonuses, while those who do poorly may be fired. Schools that do well attract students and get more financial resources, while those that do poorly may be taken over by the state. The responses of teachers and schools is not surprising: teach to the test and (in extreme cases) give students the correct answers or alter the answer sheets. Setting the target undermines the usefulness of the target.

Colleges game the system in a different way. *U.S. News & World Report*'s annual rankings of colleges and universities is carefully scrutinized by college applicants. Pomona College, where I teach, has increased its applicant pool enormously because of the free publicity *U.S. News* has given us. Students all over the United States, indeed the world, apply to Pomona because they see its high ranking in *U.S. News*. This growth of the applicant pool has, in turn, allowed Pomona to become increasingly selective (its acceptance rate is now under 7 percent), which further boosts its *U.S. News* rating. The rich get richer.

Knowing this, some colleges try to game the system. The *U.S. News* ratings criteria are readily available and colleges can play the game. Some strategies would be costly. For example, faculty salaries and the student–faculty ratio are counted by *U.S. News*, but are expensive for a college to change. Other metrics are easier to manipulate. For instance, two important yardsticks are the acceptance rate (fraction of applicants who are accepted—the lower, the better), and the yield (the fraction of accepted students who choose to come—the higher, the better). The Dean of Admissions at a small second-tier college told me how he makes his college look better by both criteria. Suppose, the college had 1,000 applicants and wants to have 200 first-year students. The 1,000 applicants can be divided into quintiles—the top 20 percent, second 20 percent, and so on. If the college were to admit the top two quintiles (a 40 percent acceptance rate), perhaps only the second quintile would choose to come (a 50 percent yield). Those in the top quintile use this college as a backup in case they don't get into anything better. When they do get into a better college, they go.

The dean's cynical strategy is to reject the top quintile because they are too good for his college, and only admit the second quintile, all of whom choose to come. Thus, he cuts his acceptance rate in half (from 40 percent to 20 percent) and doubles his yield (from 50 percent to 100 percent). The numbers aren't as cut and dried as in this stylized example, but the principle is the same. This dean rejects the best applicants, figuring that most of them wouldn't come if offered admission, and thereby lowers the acceptance rate and increases the yield, making his second-tier college look more selective and desirable. Setting the target undermines the usefulness of the target.

Another important *U.S. News* criteria is the percentage of classes with fewer than 20 students. I know one college that exploited this metric in order to game the system and rise rapidly in the *U.S. News* rankings.

Suppose, for concreteness, that the economics department offers an introductory course that is taken by 300 students each semester. The department could offer ten sections, with nine of the sections capped at 19 students and the tenth section uncapped. There will be 19 students in each of the nine sections and 129 students in the tenth section. The college can then report that 95 percent of these classes have fewer than 20 students, even though 43 percent of the students who take this course are in the section with 129 students. Setting the target undermines the usefulness of the target.

This is an inherent problem with AI programs that use our current behavior to predict our future behavior. We will change our behavior. We will wear different wristbands, visit different web sites, and change our smiles. If loan applicants are evaluated based on their bill-payment history, some people will pay their bills on time in order to get a high credit score; but that is okay because lenders want borrowers to make their loan payments on time. If computers evaluate loan applicants based on how often they charge their smartphones, some people may game the system by charging their phones more often, and that is not a sensible predictor of whether they will make their loan payments.

MAD

On September 26, 1983, Soviet early-warning satellites detected five U.S. intercontinental ballistic missiles (ICBMs) headed toward the Soviet Union. The Soviet protocol was to launch an immediate nuclear attack on the United States before incoming U.S. missiles knocked out the Soviet's response capability.

This was a game-theoretic military strategy known as mutually assured destruction (with the ironic acronym MAD). Suppose that two countries have nuclear arsenals capable of destroying each other and a credible policy of responding to a nuclear attack with an all-out nuclear assault that annihilates the country that fired first. The theoretical game-theory equilibrium is that neither country attacks the other because both nations will be destroyed.

The threat of all-out retaliation is completely credible if computers are programmed to launch a counter-attack as soon as incoming missiles are detected, and the computer programs cannot be overruled by humans. Knowing that a catastrophic response is guaranteed, neither country would dare attack the other.

This irreversible computerized response is the basis for Stanley Kubrick's satirical film *Dr. Strangelove*, with the U.S. and the Soviet Union each building Doomsday Machines that are activated by early-warning systems and cannot be stopped by human intervention.

Fortunately, when it happened in real life in 1983, there was a human involved, a Soviet lieutenant named Stanislav Petrov, who was on duty in Moscow monitoring signals from the Soviet early-warning system. When the system indicated that U.S. missiles were headed for the Soviet Union, Petrov violated protocol and did not launch a Soviet counter-attack, which surely would have been the start of a mutually assured destruction.

As Petrov hoped, it turned out to be a false alarm that was caused by an unusual reflection of sunlight off clouds that fooled the Soviet early-warning system. Imagine the consequences if the Soviets has been using a computer algorithm that could not be overridden.

This was not first (or last) nuclear near-miss. On January 25, 1995, a four-stage scientific rocket named Black Brant was launched from Norway. Brant's radar signature happened to resemble that of a U.S. Trident missile launched from a nuclear submarine. The flight path was headed towards Moscow! The Soviet early-warning system triggered a high alert, thinking that this missile might have been sent to set off a high-altitude detonation that would blind Soviet radar in preparation for a full-scale U.S. nuclear attack. Then Brant separated from its first engine and the radar signature changed to that of a multiple re-entry vehicle (MRV) armed with several nuclear warheads. Not an improvement.

Soviet computers activated Russia's nuclear briefcase, in preparation for Russian president Boris Yeltsin to authorize a nuclear response. Yeltsin had five minutes to decide whether to give final authorization. He hesitated and Brant veered away from Moscow and fell harmlessly into the ocean before five minutes had elapsed. What if there had been no human role, with a computer algorithm deciding whether or not to end life as we know it?

It is easy to imagine a future when wars are conducted entirely by computerized equipment—fully automated planes, drones, tanks, and robots firing missiles, bombs, lasers, and bullets. For individual targets, an AI system would be given a description of a person, place, or thing and instructed to attack when its visual recognition software found a match. Military robots are known as lethal autonomous weapons (LAWs).

South Korea has Samsung SGR-A1 LAWs guarding the Korean Demilitarized Zone, which is the most militarized border in the world. Very little is known for certain about these weapons, but reports are that they can detect movement up to two miles away, identify and videotape targets, send warnings, and fire bullets or grenades. Samsung says that humans must approve the decision to use the SGR machine guns or grenade launchers, but some independent outside groups believe that the systems will fire automatically unless overridden by humans.

Presumably, future AI weapon systems will have built-in loss functions, like in checkers or chess, that assess the possible consequences of alternative actions before proceeding. The loss functions will have to be much more complex than for a board game because they will need to take into account the value of the target, the potential collateral damage (including negative publicity), and the likelihood of having opportunities to attack in other circumstances. All these factors are extremely complicated and will (or at least should) be specific to the target, location, and military situation.

It is not just in the military. When automobiles are self-driving, the AI software will sometimes have to make value judgments. Should the car swerve into dangerous traffic in order to avoid hitting a pedestrian or a school bus? I don't know the answer, but it seems that the driver should decide, not the programmer who wrote the algorithm—and perhaps neglected to consider this dilemma.

The AI algorithms that control automobiles and military equipment might calculate a probability-weighted expected value or it might be minmax—choosing the action minimizes the loss in the worst-case scenario. The rule will necessarily be subjective and debatable and, once programmed in, it will be unflinchingly inflexible. A human might think twice after noticing that a military attack will destroy a nearby religious shrine or take out a city's water supply. A computer won't care unless the programmers put that in its loss function and told the program exactly how to assess the costs.

Humans make mistakes and take actions with unintended consequences, but it seems terrifying to turn wars over to machines and see how it works out.

Conclusion

We live in an incredible period in history. The Computer Revolution may be even more life-changing than the Industrial Revolution. We can do things with computers that could never be done before, and computers can do things for us that never could be done before.

I am addicted to computers and you may be, too. But we shouldn't let our love of computers cloud our recognition of their limitations. Yes, computers know more facts than we do. Yes, computers have better memories than we do. Yes, computers can make calculations faster than we can. Yes, computers do not get tired like we do.

Robots far surpass humans at repetitive, monotonous tasks like tightening bolts, planting seeds, searching legal documents, and accepting bank deposits and dispensing cash. Computers can recognize objects, draw pictures, drive cars. You can surely think of a dozen other impressive—even superhuman—computer feats.

It is tempting to think that because computers can do some things extremely well, they must be highly intelligent. However, being useful for specific tasks is very different from having a general intelligence that applies the lessons learned and skills required for one task to more complex tasks or to completely different tasks. With true intelligence, skills are portable.

Computers are great and getting better, but computer algorithms are still designed to have the very narrow capabilities needed to perform well-defined chores, not the general intelligence needed to deal with unfamiliar situations by assessing what is happening, why it is happening, and what the consequences are of taking action. Humans can apply general knowledge to specific situations and use specific situations to improve their general knowledge. Computers today cannot.

Artificial intelligence is not at all like the real intelligence that comes from human brains. Computers do not know what words mean because computers do not experience the world the way we do. They do not even know what the real world is. Computers do not have the common sense or wisdom that humans accumulate by living life. Computers cannot formulate persuasive theories. Computers cannot do inductive reasoning or make long-run plans. Computers do not have the emotions, feelings, and inspiration that are needed to write a compelling poem, novel, or movie script.

There may come a time when computers possess real human-like intelligence, but it won't be because computer memories grow larger and processing speeds get faster. It is not a question of quantitative improvements. It is a question of a qualitatively different approach—of finding ways for computers to acquire a general intelligence that can be broadly applied in a flexible manner to unfamiliar situations.

To be clear, I am not criticizing computer scientists. They are terrifically intelligent and extraordinarily hard working. What computer scientists have done is very difficult and extremely useful. What more can be done is harder still, orders of magnitude harder.

Mimicking the human mind is a daunting task with no guarantee of success. There have been some legendary exceptions, like AT&T's Bell Labs, Lockheed Martin's Skunk Works, and Xerox's Parc, but few companies are willing to support intellectually interesting research that does not have a short-term payoff. It is more appealing to make something that is useful and immediately profitable.

I don't know how long it will take to develop computers that have a general intelligence that rivals humans. I suspect that it will take decades. I am certain that people who claim that it has already happened are wrong, and I don't trust people who give specific dates, like 2029. In the meantime, please be skeptical of far-fetched science fiction scenarios and please be wary of businesses hyping AI products.

Data mining Big Data is all the rage, but data mining is artificial and it isn't intelligent. When statistical models analyze a large number of potential explanatory variables, the number of possible relationships becomes astonishingly large. With a thousand possible explanatory variables in a multiple regression model, there are nearly a trillion trillion possible combinations of ten input variables. There are more than a billion trillion trillion ten-variable combinations when there are ten thousand possible explanatory variables, and unimaginably more when there are a million possible explanatory variables.

If many potential variables are considered, even if all of them are just random noise, some combinations are bound to be highly correlated with whatever it is we are trying to predict: cancer, credit risk, job suitability. There will occasionally be a true knowledge discovery, but the larger the number of explanatory variables considered, the more likely it is that a discovered relationship will be coincidental and transitory.

Statistical evidence is not sufficient to distinguish between real knowledge and bogus knowledge. Only logic, wisdom, and common sense can do that. Computers cannot assess whether things are truly related or just coincidentally correlated because computers do not understand data in any meaningful way. Numbers are just numbers. Computers do not have the human judgment needed to tell the difference between good data and bad data. Computers do not have the human intelligence needed to distinguish between statistical patterns that make sense and those that are spurious. Computers today can pass the Turing Test, but not the Smith Test. The situation is exacerbated if the discovered patterns are concealed inside black boxes that make the models inscrutable. Then no one knows why a computer algorithm concluded that this stock should be purchased, this job applicant should be rejected, this patient should be given this medication, this prisoner should be denied parole, this building should be bombed.

In the age of Big Data, the real danger is not that computers are smarter than us, but that we *think* computers are smarter than us and therefore trust computers to make important decisions for us. We should not be intimidated into thinking that computers are infallible, that data mining is knowledge discovery, that black boxes should to be trusted. Let's trust ourselves to judge whether statistical patterns make sense and are therefore potentially useful, or are merely coincidental and therefore fleeting and useless.

Human reasoning is fundamentally different from artificial intelligence, which is why it is needed more than ever.

BIBLIOGRAPHY

1938 New England hurricane. (n.d.). In Wikipedia. Retrieved April 29, 2015, from: http://en.wikipedia.org/wiki/1938_New_England_hurricane#cite_note-3.

Alba, Davey. 2011. How Siri responds to questions about women's health, sex, and drugs, *Laptop*, December 2.

Alexander, Harriet. 2013. 'Killer Robots' could be outlawed, *The Telegraph*, November 14.

Allan, Nicole, Thompson, Derek. 2013. The myth of the student-loan crisis, *The Atlantic*, March.

Anderson, Chris. 2008. The end of theory, will the data deluge make the scientific method obsolete?, *Wired*, June 23.

Andrew, Elise. undated, AI trying to design inspirational posters goes horribly and hilariously wrong, IFLScience, Available from: http://www.livemint.com/Technology/VXCMw0Vfilaw0aIInD1v2O/When-artificial-intelligence-goes-wrong.html.

Angwin, Julia, Larson, Jeff, Mattu, Surya, Kirchner, Lauren. 2016. Machine bias, *ProPublica*, May 23. Available from: https://www.propublica.org/article/machine-bias-risk-assessments-in-criminal-sentencing.

Angwin, Julia, Scheiber, Noam, Tobin, Ariana. 2017. Machine bias: dozens of companies are using facebook to exclude older workers from job ads, ProPublica December 20., Available from: https://www.propublica.org/article/facebook-ads-age-discrimination-targeting.

Anonymous. 2015. A long way from dismal: economics evolves. *The Economist*, 414 (8920), 8.

Baum, Gabrielle, Smith, Gary. 2015. Great companies: looking for success secrets in all the wrong places, *Journal of Investing*, 24 (3), 61–72.

Bem, DJ. 2011. Feeling the future: experimental evidence for anomalous retroactive influences on cognition and affect, *Journal of Personality and Social Psychology*, 100 (3), 407–25.

Berk, R. 2013. Algorithmic criminology, *Security Informatics*, 2: 5. https://doi.org/10.1186/2190-8532-2-5.

Bolen, Johan, Mao, Huina, Zeng, Xiaojun, 2011. Twitter mood predicts the stock market, *Journal of Computational Science*, 2 (1), 1–8.

Boyd, Danah, Crawford, Kate. 2011. Six provocations for big data. A Decade in Internet Time: Symposium on the Dynamics of the Internet and Society, September 21, 2011.

Brennan-Marquez, Kiel. 2017. Plausible cause: explanatory standards in the age of powerful machines, *Vanderbilt Law Review*, 70 (4), 1249–1301.

Brodski, A, Paasch, GF, Helbling, S, Wibral, M. 2015. The faces of predictive coding. *Journal of Neuroscience*, 35 (24): 8997.

Calude, Cristian S, Longo, Giuseppe. 2016. The deluge of spurious correlations in big data, *Foundations of Science*, 22, 595–612. https://doi.org/10.1007/s10699-016-9489-4.

Cape Cod Times, July 7, 1983.

Caruso, EM, Vohs, KD, Baxter, B, Waytz, A. 2013. Mere exposure to money increases endorsement of free-market systems and social inequality. *Journal of Experimental Psychology: General*, 142, 301–306. http://dx.doi.org/10.1037/a0029288.

Chappell, Bill. 2015. Winner of French Scrabble title does not speak French, The Two-Way: Breaking News From NPR.

Chatfield, Chris. 1995. Model uncertainty, data mining and statistical inference, *Journal of the Royal Statistical Society A* 158, 419–466.

Chollet, Francois. 2017. *Deep Learning With Python*. Manning Publications.

Christensen, B, Christensen, S. 2014. Are female hurricanes really deadlier than male hurricanes? *Proceeding of the National Academy of Sciences USA*, 111 (34) E3497–98.

Coase R, 1988. How should economists choose?, in *Ideas, Their Origins and Their Consequences: Lectures to Commemorate the Life and Work of G. Warren Nutter*, American Enterprise Institute for Public Policy Research.

Cohn, Carolyn, 2016, Facebook stymies Admiral's plans to use social media data to price insurance premiums, *Reuters*, November 2. Available from: https://www.reuters.com/article/us-insurance-admiral-facebook/facebook-stymies-admirals-plans-to-use-social-media-data-to-price-insurance-premiums-idUSKBN12X1WP.

Dalal, SR, Fowlkes, EB, Hoadley, B. 1989. Risk analysis of the space shuttle: pre-Challenger prediction of failure, *Journal of the American Statistical Association*. 84 (408), 945–57.

Davis, Ernest, 2014. The technological singularity: the singularity and the state of the art in artificial intelligence. *Ubiquity*, Association for Computing Machinery. October. Available from: http://ubiquity.acm.org/article.cfm?id=2667640.

Devlin, Hannah, 2015. Rise of the robots: how long do we have until they take our jobs?, *The Guardian*, February 4.

Diamond, Jared, 1989. How cats survive falls from New York skysrcapers, *Natural History*, 98, 20–26.

Duhigg, Charles, 2012. How companies learn your secrets, *The New York Times Magazine*. February 16.

Effects of Hurricane Sandy in New York. (n.d.) In Wikipedia. Retrieved April 29, 2015, from: http://en.wikipedia.org/wiki/Effects_of_Hurricane_Sandy_in_New_York.

Evtimov, I, Eykholt, K, Fernandes, EKohno, T, Li, B, Prakash, A. et al. 2017. Robust physical-world attacks on deep learning models. <https://arxiv.org/abs/1707.08945>

Fayyad, Usama, Piatetsky-Shapiro, Gregory, Smyth, Padhraic. 1996. From data mining to knowledge discovery in databases, AI Magazine, 17 (3), 37–54.

Galak, J, LeBoeuf, RA, Nelson, LD, & Simmons, JP. 2012. Correcting the past: failures to replicate Psi. *Journal of Personality and Social Psychology*, 103(6), 933–48.

Garber, Megan, 2016. When algorithms take the stand, *The Atlantic*, June 30.

Gasiorowska, Agata, Chaplin, Lan Nguyan, Zaleskiewicz, Tomasz, Wygrab, Sandra, Vohs, Kathleen D. 2016. Money cues increase agency and decrease prosociality among children: early signs of market-mode behaviors, *Psychological Science*, 27(3), 331–44.

Gebeloff, Robert, Dewan, Shaila. 2012. Measuring the top 1% by wealth, not income, *New York Times*, January 17.

Gemini. Is more education better? 2014. *DegreeCouncil*, April 18. Available from: http://degreecouncil.org/2014/is-more-education-better/.

Goldmacher, Shane. 2016. Hillary Clinton's "Invisible Guiding Hand", *Politico*, September 7.

Goldman, William, 1983. Adventures in the screen trade, Los Angeles: Warner Books.

Greene, Lana. 2017. Beyond Babel: The limits of computer translations, *The Economist*, January 7. 422 (9002), 7.

Herkewitz, William. 2014. Why Watson and Siri are not real AI, *Popular Mechanics*, February 10.

Hirshleifer, D, Shumway, T. 2003. Good day sunshine: Stock returns and the weather. *Journal of Finance*, 58(3) 1009–32.

Hou, Kewei, Xue, Chen, Zhang Lu. 2017. Replicating anomalies, *NBER Working Paper No. 23394*, May.

Hoerl, Arthur E, Kennard, Robert W. 1970a. Ridge regression: Biased estimation for nonorthogonal problems. *Technometrics*, 12, 55–67.

Hoerl, Arthur E, and Kennard, Robert W. 1970b. Ridge regression: Applications to nonorthogonal problems. *Technometrics*, 12, 69–82.

Hofstadter, Douglas. 1979. *Gödel, Escher, Bach: An Eternal Golden Braid*, New York: Basic Books.

Hofstadter, Douglas, Sander, Emmanuel. 2013, *Surfaces and Essences: Analogy as the Fuel and Fire of Thinking*, New York: Basic Books.

Hornby, Nick, 1992. Fever Pitch, London: Victor Gollancz Ltd., p. 163.

Hurricane Allen. (n.d.). In Wikipedia. Retrieved April 29, 2015, from: http://en.wikipedia.org/wiki/Hurricane_Allen#cite_note-1983_Deadly-18.

Hvistendahl, Mara. You are a number, *Wired*, January 2018, 48–59.

Ip, Greg. 2017. We survived spreadsheets, and we'll survive AI, *The Wall Street Journal*, August 3.

Issenberg, Sasha. 2012. How Obama used big data to rally voters, Part 1, *MIT Technology Review*.

Johnson P. 2013. 75 years after the Hurricane of 1938, the science of storm tracking improved significantly. *MassLive*, September 14. Available from: http://www.masslive.com/news/index.ssf/2013/09/75_years_after_the_hurricane_o.html, Retrieved April 29, 2015.

Jung, K, Shavitt, S, Viswanathan, M, Hilbe, JM. 2014. Female hurricanes are deadlier than male hurricanes. *Proceedings of the National Academy of Sciences. USA* 111 (24), 8782–7.

Kendall, MG. 1965. *A Course in Multivariate Statistical Anaalysis*. Third Edition, London: Griffin.

Khomami, Nadia. 2014. 2029: the year when robots will have the power to outsmart their makers, *The Guardian*, February 22.

Knight, Will. 2016. Will AI-powered hedge funds outsmart the market?, *MIT Technology Review*, February 4.

Knight, Will. 2017. The financial world wants to open AI's black boxes, *MIT Technology Review*, April 13.

Knight, Will. 2017. There's a dark secret at the heart of artificial intelligence: no one really understands how it works, *MIT Technology Review*, April 11.

Knight, Will. 2017. Alpha Zero's alien chess shows the power, and the peculiarity, of AI, *MIT Technology Review*, December 8.

Labi, Nadia, 2012. Misfortune Teller, *The Atlantic*, January/February 2012.

Lambrecht, Anja, Tucker, Catherine E. Algorithmic bias? An empirical study into apparent gender-based discrimination in the display of STEM career ads (November 30, 2017). Available at SSRN: https://ssrn.com/abstract=2852260 or http://dx.doi.org/10.2139/ssrn.2852260.

Lazer, David, Kennedy, Ryan, King, Gary, Vespignani, Alessandro. 2014. The parable of Google flu: traps in big data analysis, *Science*, 343 (6176), 1203–5.

LeClaire, Jennifer. 2015. Moore's Law turns 50, creator says it won't go on forever, *Newsfactor*, May 12.

Leswing, Kif. 2016. 21 of the funniest responses you'll get from Siri, *Business Insider*, March 28.

Lewis-Kraus, Gideon. 2016. The great AI awakening, *The New York Times Magazine*, December 14.

Liptak, Adam. 2017. Sent to prison by a software program's secret algorithms, *New York Times*, May 1.

Loftis, Leslie. 2016. How Ada let Hillary down, Arc Digital, December 13.

Madrigal, Alexis. 2013. Your job, their data: the most important untold story about the future, *Atlantic*, November 21.

Maley, S. 2014. Statistics show no evidence of gender bias in the public's hurricane preparedness. *Proceedings of the National Academy of Sciences,* 111 (37) E3834. USA. http://dx.doi.org/10.1073/pnas.1413079111.

Malter, D. 2014. Female hurricanes are not deadlier than male hurricanes. *Proceeding of the National Academy of Sciences. USA,* 111 (34) E3496. http://dx.doi.org/10.1073/pnas.1411428111.

Marquardt, Donald W. 1970. Generalized inverses, ridge regression, biased linear estimation. *Technometrics*, 12, 591–612.

Massy, William F. 1965. Principal components in exploratory statistical research. *Journal of the American Statististical Association*, 60, 234–56.

McLean, R. David, Pontiff Jeffrey. 2016. Does academic research destroy stock return predictability?, *Journal of Finance*, 71 (1), 1540–6261.

Metz, Cade. 2016. Trump's win isn't the death of data—it was flawed all along, *Wired*, November 9.

Minkel, J R. 2007. Computers solve checkers—it's a draw, *Scientific American*, July 19.

National Oceanic and Atmospheric Administration. 2012. Predicting hurricanes: times have changed. Available from: http://celebrating200years.noaa.gov/magazine/devast_hurricane/welcome.html.

Nguyen, Anh, Yosinski, Jason, Clune Jeff. 2015. Deep neural networks are easily fooled: high confidence predictions for unrecognizable images, *Proceedings of the IEEE Conference on Computer Vision and Pattern Recognition*. Available from: https://arxiv.org/abs/1412.1897v4.

Otis, Ginger Adams. 2014. Married people have less heart problems than those who are single, divorced: study, *Daily News*, March 28.

Panesar, Nirmal, Graham, Colin. 2012. Does the death rate of Hong Kong Chinese change during the lunar ghost month? *Emergency Medicine Journal* 29, 319–21.

Panesar, Nirmal S, Chan, Noel CY, Li, Shi N, Lo, Joyce KY, Wong, Vivien WY, Yang, Isaac B, Yip, Emily KY. 2003. Is four a deadly number for the Chinese?, *Medical Journal of Australia*, 179 (11): 656–8.

Peck, Don. 2013. They're Watching You at Work, *Atlantic*, December.

Pektar, Sofia. 2017. Robots will wipe out humans and take over in just a few centuries warns Royal astronomer, *Sunday Express*, April 4.

Poon, Linda. 2014. Sit more, and you're more likely to be disabled after age 60, *NPR*, February 19.

Ruddick, Graham. 2016. Admiral to price car insurance based on Facebook posts, *The Guardian*, November 1.

Reese, Hope. 2016. Why Microsoft's 'Tay' AI bot went wrong, *TechRepublic*, Available at: http://www.techrepublic.com/article/why-microsofts-tay-ai-bot-went-wrong, March 24.

Reilly, Kevin. 2016. Two of the smartest people in the world on what will happen to our brains and everything else, *Business Insider*, January 18.

Ritchie, Stuart J, Wiseman, Richard, French, Christopher C, Gilbert, Sam. 2012. Failing the future: three unsuccessful attempts to replicate bem's 'retroactive facilitation of recall' effect, *PLoS ONE*. 7 (3): e33423.

Robbins, Bruce, Ross, Andrew. 2000. Response: Mystery science theatre, in: *The Sokal Hoax: The Sham that Shook the Academy*, edited by Alan D. Sokal, University of Nebraska Press, pp. 54–8.

Rudgard, Olivia. 2016. Admiral to use Facebook profile to determine insurance premium, *The Telegraph*, November 2.

Rutkin, Aviva Hope. 2017. The tiny changes that can cause AI to fail, *BBC Future*, April 17.

Sharif, Mahmood, Bhagavatula, Sruti, Bauer, Lujo, Reiter, Michael K. 2016. Accessorize to a Crime: Real and Stealthy Attacks on State-of-the-Art Face Recognition, *Proceedings of the 2016 ACM SIGSAC Conference on Computer and Communications Security*, 1528–40.

Siegel, Eric, 2016. How Hillary's campaign is (almost certainly) using big data, *Scientific American*, September 16.

Simonite, Tom. 2017. How to upgrade judges with machine learning, *MIT Technology Review*, March 6.

Singh, Angad. 2015. The French Scrabble champion who speaks no French, CNN, July 22.

Smith, Gary, Campbell, Frank, 1980. A critique of some ridge regression methods, *Journal of the American Statistical Association*, with discussion and rejoinder, 75 (369), 74–81.

Smith, Gary, 1980. An example of ridge regression difficulties, *The Canadian Journal of Statistics*, 8 (2) 217–25.

Smith, Gary, 2011. Birth month is not related to suicide among major league baseball players, *Perceptual and Motor Skills*, 112 (1), 55–60.

Smith, Gary, 2012. Do people whose names begin with d really die young?, *Death Studies*, 36 (2), 182–9.

Smith, Gary, Zurhellen, Michael. 2015. Sunny upside? The relationship between sunshine and stock market returns, *Review of Economic Analysis*, 7, 173–83.

Smith, Gary. 2014. *Standard Deviations: Flawed Assumptions, Tortured Data, and Other Ways to Lie With Statistics*, New York: Overlook.

Smith, Stacy, Allan, Ananda, Greenlaw, Nicola, Finlay, Sian, Isles, Chris. 2013. Emergency medical admissions, deaths at weekends and the public holiday effect. Cohort study, *Emergency Medicine Journal*, 31(1):30–4.

Sokal, A. 1996. Transgressing the boundaries: Towards a transformative hermeneutics of quantum gravity. *Social Text*, 46/47, 217–52.

Somers, James. 2013. The man who would teach machines to think, *The Atlantic*, November.

Stanley, Jason, and Vesla Waever, 2014. Is the United States a Racial Democracy?, *New York Times*, Online January 12.

Su, Jiawei, Vargas, Danilo Vasconcellos, Kouichi, Sakurai. 2017. One pixel attack for fooling deep neural networks, November 2017. Available from: https://arxiv.org/abs/1710.08864.

Szegedy, Christian, Zaremba, Wojciech, Sutskever, Ilya, Bruna, Joan, Erhan, Dumitru, Goodfellow, Ian, Fergus, Rob. 2014. Intriguing properties of neural networks, Google, February 19. Available from: https://www.researchgate.net/publication/259440613_Intriguing_properties_of_neural_networks

Tal, Aner, Wansink, Brian. 2014. Blinded by science: trivial scientific information can increase our sense of trust in products, *Public Understanding of Science*. 25, 117–25.

Tashea, Jason. 2017. Courts are using AI to sentence criminals. that must stop now. *Wired*, April 17.

Tatem, Andrew J, Guerra, Carlos A, Atkinson, Peter M, Hay, Simon I. 2004. Athletics: momentous sprint at the 2156 Olympics?, *Nature*, 431, 525.

Todd, Chuck, Dann, Carrie. 2017. How big data broke American politics, *NBC News*, March 14.

Traxler MJ, Foss DJ, Podali R, Zirnstein M. 2012. Feeling the past: the absence of experimental evidence for anomalous retroactive influences on text processing, *Memory and Cognition*. 40(8), 1366–72.

Vadillo, Michael A, Hardwicke, Tom E, Shanks, David R., 2016. Selection bias, vote counting, and money-priming effects: a comment on Rohrer, Pashler, and Harris (2015) and Vohs (2015), *Journal of Experimental Psychology. General*, 145(5), 655–63.

Vohs KD, Mead, NL, Goode MR. 2006. The psychological consequences of money. *Science*, 314, 1154–6. http://dx.doi.org/10.1126/science.1132491.

Vorhees, William. 2016. Has AI gone too far? Automated inference of criminality using face images, *Data Science Central*, November 29.

Wagner, John. 2016. Clinton's data-driven campaign relied heavily on an algorithm named Ada. What didn't she see?, *Washington Post*, November 9.

Walsh, Michael. 2017. UPDATE: A.I. inspirational poster generator suffers existential breakdown. Nerdist, June 27. http://nerdist.com/a-i-generates-the-ridiculous-inspirational-posters-that-we-need-right-now/

Weaver, John Frank. 2017. Artificial intelligence owes you an explanation. *Slate*, May 8.

Wiecki, Thomas, Campbell, Andrew, Lent, Justin, Stauth, Jessica. 2016. All that glitters is not gold: comparing backtest and out-of-sample performance on a large cohort of trading algorithms, *Journal of Investing*, 25 (3), 69–80.

Willer, Robb. 2004. The Intelligibility of Unintelligible Texts. Master's thesis. Cornell University, Department of Sociology.

Willsher, Kim. 2015. The French Scrabble champion who doesn't speak French, *The Guardian*, July 21.

Winograd, Terry. 1972. Understanding natural language. *Cognitive Psychology*, 3, 1–191.

Wolfson, Sam. 2016. What was really going on with this insurance company basing premiums on your facebook posts?, *Vice*, November 2.

Woollaston, Jennifer. 2016. Admiral's firstcarquote may breach Facebook policy by using profile data for quotes, *Wired UK*, November 2.

Wu, Xiaolin, Zhang, Xi. 2016. Automated inference on criminality using face images, Shanghai Jiao Tong University, November 21. Available at: https://arxiv.org/abs/1611.04135v1.

Wu, Xiaolin, Zhang Xi. 2017. Responses to critiques on machine learning of criminality perceptions, Shanghai Jiao Tong University, May 26. Available at: https://arxiv.org/abs/1611.04135v3.

Yuan, Li. 2017. Want a loan in China? Keep your phone charged, *The Wall Street Journal*, April 6.

Yudkowsky, Eliezer. 2008. Artificial intelligence as a positive and negative factor in global risk. In *Global Catastrophic Risks*, edited by Nick Bostrom and Milan M. Ćirković, New York: Oxford University Press, 308–45.

Zeki, Semir, Romaya, John Paul, Benincasa, Dionigi MT, Atiyah Michael F. 2014. The experience of mathematical beauty and its neural correlates, *Frontiers in Human Neuroscience*, February 13. 8, article 68.

Zuckerman, Gregory, Hope, Bradley. 2017. The quants run Wall Street now, *The Wall Street Journal*, May 21.

INDEX